Bearings and Seals

PUBLISHER

Fundamentals of Service (FOS) is a series of manuals created by Deere & Company. Each book in the series was conceived, researched, outlined, edited, and published by Deere & Company. Authors are selected to provide a basic technical manuscript which is edited and rewritten by editors.

PUBLISHER: DEERE & COMPANY SERVICE PUBLICATIONS. Dept. FOS/FMO, John Deere Road, Moline, Illinois 61265-8098; DEPT. MANAGER: Alton E. Miller.

FUNDAMENTAL WRITING SERVICES EDITORIAL STAFF
Managing Editor: Louis R. Hathaway
Editor: John E. Kuhar
Publisher: Lori J. Lees
Promotions: Cindy S. Calloway

TO THE READER

PURPOSE OF THIS MANUAL

This manual can easily help train a reader to understand bearings and seals better, and service them with speed and skill. Starting with the question "how it works," the book builds up to "why it fails" and "what to do about it." This manual is also a reference for the service technician who wants to refresh his or her memory on bearings and seals. It is written in a clear form with many illustrations.

 CAUTION: Observe safety messages and information in this manual.

WHAT IS "FUNDAMENTALS OF SERVICE"?

This manual is part of a series of texts and visuals called "Fundamentals of Service," or "FOS." These materials give basic information on power mechanics for teachers as well as shop service technicians and the lay reader. All kinds of equipment are covered; both automotive and off-the-road with emphasis on theory of operation, diagnosis, and repair.

OTHER MANUALS IN THIS SERIES

Other manuals in the FOS series are:

- **Hydraulics**
- **Engines**
- **Electrical Systems**
- **Power Trains**
- **Shop Tools**
- **Air Conditioning**
- **Tires and Tracks**
- **Belts and Chains**
- **Welding**
- **Mowing and Spraying Equipment**
- **Fuels, Lubricants, and Coolants**
- **Fiberglass and Plastics**
- **Fasteners**
- **Identification of Parts Failures**
- **A Glossary of Technical Terms**

Each manual is backed up by a set of 35 mm color slides for classroom use. Transparency masters, instructor's guides, and student guides are also available for the first four subjects.

FOR MORE INFORMATION

Write for a free *Catalog of Educational Materials*. Send your request to:

John Deere Service Publications
Dept. FOS/FMO
John Deere Road
Moline, Illinois 61265-8098

 We have a long-range interest in good service

ACKNOWLEDGEMENTS

John Deere gratefully acknowledges help from the following groups: **For bearings:** American Oil Co.; Bendix Corp.; Bohn Aluminum & Brass Co.; Clevite Corp.; The Fafnir Bearing Co.; Federal-Mogul Corp.; FMC Corp.; New Departure — Hyatt Bearings; SKF Industries; The Timken Roller Bearing Co.; The Torrington Co. **For seals:** Acadia Synthetic; Aeroquip Corp.; Apex Molded Products Co.; Borg-Warner Corp.; Chicago Rawhide Co.; Dana Corp.; Garlock, Inc.; B. F. Goodrich Co.; Loctite Corp.; Minnesota Rubber Co.; Parker Seal Co.; Permatex Co.; Raybestos-Manhattan, Inc.; Sealed Power Corp.

CONTENTS

Litho in U.S.A.

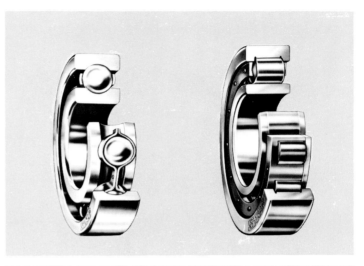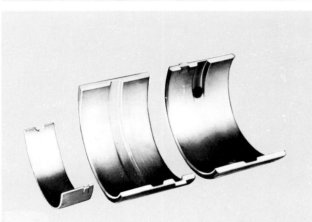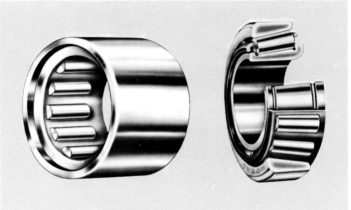

Fig. 1 — Every Type Of Bearing Is Used In Modern Machines

INTRODUCTION

Turning gears, wheels, and shafts are held in place and protected from friction and wear by bearings.

Bearings perform these functions:

1. **Support the moving part**
2. **Reduce friction**
3. **Reduce wear**
4. **Provide a replaceable wear surface**

Bearings SUPPORT MOVING PARTS by carrying radial and axial loads while holding the parts in alignment.

Bearings REDUCE FRICTION by providing either *sliding* contact (plain bearings) or *rolling* contact (anti-friction bearings). Bearings further reduce friction by using a lubricant.

By reducing friction, bearings REDUCE WEAR. They also provide a REPLACEABLE WEAR SURFACE which is more economical to replace than the gear, wheel, or shaft which the bearing supports.

PLAIN BEARINGS   SPLIT BEARING

PLAIN BEARINGS
(BUSHINGS)

X2889

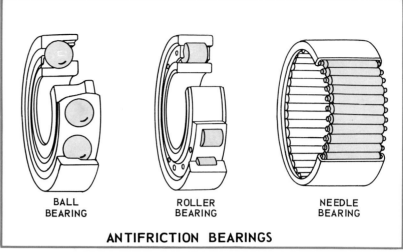

BALL BEARING   ROLLER BEARING   NEEDLE BEARING

ANTIFRICTION BEARINGS

Fig. 2 — Two Basic Types Of Bearings

TYPES OF BEARINGS

There are two basic types of bearings:

• **Plain bearings (bushings)**

• **Antifriction bearings**

PLAIN BEARINGS (Fig. 2) provide a *sliding* contact between the mating surfaces. They are also called "bushings" or "journal bearings."

ANTIFRICTION BEARINGS (Fig. 2) give a *rolling* contact between mating surfaces. They do this by using balls or rollers.

LOAD FORCES ON BEARINGS

Because bearings may be subjected to *axial* or *radial* loads (Fig. 3), they are designed for specific applications.

Radial bearings are used where only radial (sideways) loads are encountered. Thrust bearings are used where only axial (endways) loads are found. Some bearings are made to withstand both radial and axial loads.

Thrust loads are also carried by tapered roller bearings and some straight roller ball bearings. Due to their design, a radial load is converted to a thrust load in these bearings.

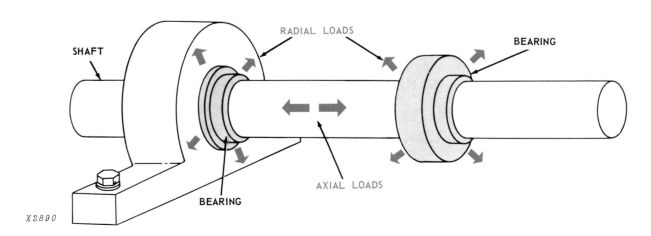

SHAFT   RADIAL LOADS   BEARING

AXIAL LOADS

BEARING

X2890

Fig. 3 — Load Forces Acting On Bearings

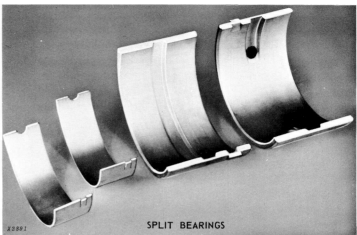

Fig. 4 — Typical Plain Bearings

ADVANTAGES AND DISADVANTAGES OF BEARINGS

Both plain and antifriction bearings have advantages and disadvantages. Let's compare them.

PLAIN BEARINGS (BUSHINGS)

Advantages

1. Require little space

2. Low cost

3. Quiet operation

4. Rigid construction

5. Ease of installation

Disadvantages

1. High friction operation

2. Require continuous lubrication because they cannot be packed with lubricant

ANTIFRICTION BEARINGS

Advantages

1. Low friction operation

2. Can be packed with lubricant to reduce frequency of lubrication

3. More versatile because there are many designs

Disadvantages

1. Require more space than plain bearings

2. Noisier operation

3. Higher cost than plain bearings

4. Less rigid than plain bearings

Now that we have described bearings in general, let's examine each basic type in more detail: First, plain bearings.

PLAIN BEARINGS (BUSHINGS)

Plain bearings, as shown in Fig. 4, are often called *bushings*. Since "bearing" and "bushing" are synonymous terms, we will refer to all such designs as *bearings*.

Plain bearings take less space and usually cost less than antifriction bearings. If frequent service is necessary and the location is accessible, plain bearings may be used.

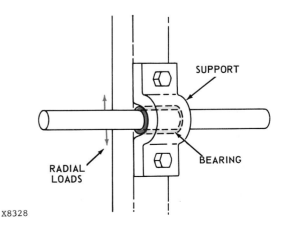

Fig. 5 — Bearing Designed For Radial Loads Only

However, antifriction bearings are normally used where space is available, the location is not readily accessible, loads are heavy, speeds are high, less service is required, and cost is not the critical factor.

LOADS ON PLAIN BEARINGS

Before describing plain bearings, let's compare bearings designed for *radial* loads and *thrust* loads. In Fig. 5 the bearing is supporting radial loads only.

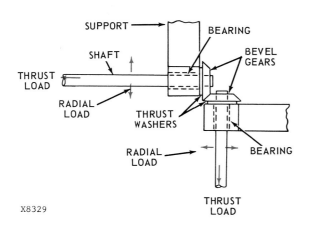

X8329

Fig. 6 — Thrust Washers Used With Radial-Load-Type Bearings

Radial load bearings may be used with thrust washers, as shown in Fig. 6, when both thrust and radial loads are present. A thrust washer is a plain thrust bearing.

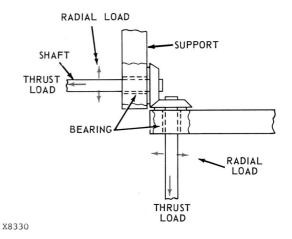

X8330

Fig. 7 — Bearings Designed For Both Radial And Thrust Loads

Bearings that support radial loads as well as axial loads are shown in Fig. 7.

TYPES OF PLAIN BEARINGS

Fig. 8 below shows some typical plain bearings. They are described on the next page.

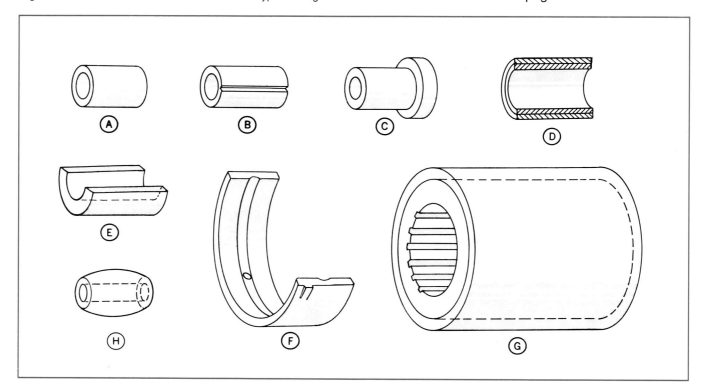

Fig. 8 — Types Of Plain Bearings

A. **Solid bearing or sleeve**. *These bearings are usually made of copper, brass, bronze, or plastic. They only support radial loads.*

B. **Split bearings** *are initially formed into a strip then rolled into final shape. Split bearings are usually made of copper, bronze, or steel. They only support radial loads.*

C. **Thrust bearings** *can carry both radial and thrust loads. They are usually made of bronze or plastic.*

D. *This type of* **split bearing** *is normally constructed with a steel back lined with bronze, plastic, or babbit. It supports radial loads only.*

E. *This is a* **split bearing** *that may be constructed of wood, plastic, rubber, sintered metal or cast iron. This bearing also supports radial loads only.*

F. *This type of* **split bearing** *is normally used for engine crankshaft bearings. It is usually made with a steel or bronze back with a softer babbit lining or of an aluminum alloy. However, there are linings such as copper-lead, tin, or silver, which will be discussed later. Variations of this bearing will support both radial and thrust loads.*

G. *This* **solid bearing** *has a fluted rubber structure supported by a rubber shell. The flutes in the rubber form a series of passages that allow the lubricant, normally water, to flow through the bearing. This bearing supports radial loads only.*

H. *This* **self-aligning bearing** *has a spherical outer surface so that it can align itself with the shaft. It may be mounted in a spherical housing or in a flexible mounting such as rubber. It often contains a groove to prevent bearing rotation.*

MOUNTINGS FOR PLAIN BEARINGS

Plain bearings are usually installed in their supports with a *press fit* or a shrink fit. This fit prevents the bearing from rotating within its mounting, assures good heat transfer which is important to bearing life, and assures good concentricity of the bearing.

The running clearance is between the inside surface of the bearing and the O.D. of the shaft.

Some plain bearings are mounted with seals which retain the lubricant and exclude dirt. (See the section on "Seals" later in the manual.)

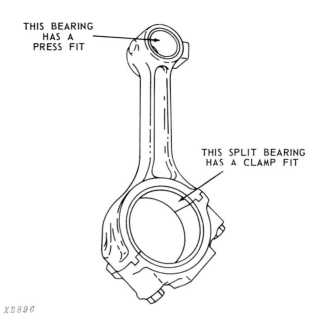

Fig. 9 — Two Types Of Bearing Installations

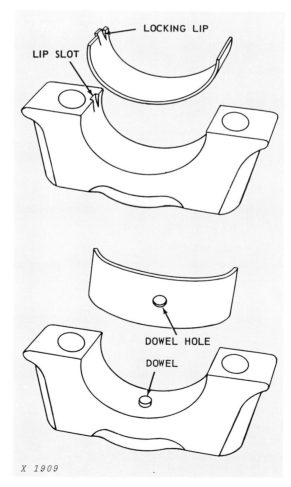

Fig. 10 — Two Types Of Bearing Locks

In Fig. 9 you can see two types of bearing installations in an engine connecting rod.

The upper eye of the connecting rod has a *press-fit* bearing which in this case requires reaming after installation. This reaming is required to size the bearing to the piston pin.

On the other end of the connecting rod is a split bearing that is *clamped* together.

Split bearings are usually locked in place to prevent rotation with the shaft. Two methods of locking are shown in Fig. 10.

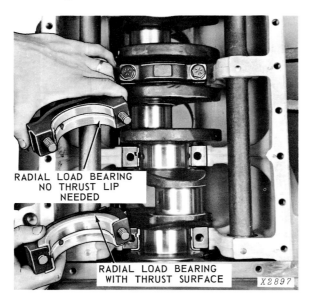

RADIAL LOAD BEARING NO THRUST LIP NEEDED

RADIAL LOAD BEARING WITH THRUST SURFACE

X2897

Fig. 11 — Radial And Thrust Split Bearings

While we are talking about split bearings, notice the difference between radial and thrust split bearings as shown in Fig. 11.

Some bearings have a running clearance between the shaft and bearing in addition to clearance between the bearing O.D. and its housing. These bearings are termed *floating bearings* and are used in such devices as turbochargers where the speed is very high (80,000 rpm and up) and the load is light. They are protected by a "cushion" of lubricant.

MATERIALS IN PLAIN BEARINGS

Bearings are constructed of a variety of materials. Some of these are: Wood, plastic, cast iron, chilled cast iron, soft steel, hardened steel, copper, brass, bronze, copper-lead, aluminum, babbit, sintered metal, iron, brass, copper, graphite, phenolic material, and nylon.

The material used in a bearing depends on the space available, the speed the shaft turns, the amount of load it carries, the type of lubricant used, and, in many cases, the cost of the installation.

1. Wood bearings are sometimes used on large slow-moving shafts, such as conveyors, which are subjected to excessive amounts of dirt. They are not costly, require a minimum of attention, have relatively long life, and are easily replaced.

2. Cast iron and chilled (hardened) cast iron bearings are usually used on slow-moving shafts that operate in excessively dirty conditions. Since these bearings are resistant to the wear caused by dirt, they are used on such implements as disk harrows, plows, and other slow-moving machines.

3. Soft steel bearings usually support shafts having little movement and where support is the main objective.

Hard steel bearings are used to support large slow-moving shafts or as bearings in slow-moving wheels and the like.

4. Copper, brass, and bronze bearings are widely used to support shafts and gears in machines. Normally, the material is determined by the application; each metal has different wearing characteristics. Bronze is probably the most commonly used.

5. Copper-lead bearings are an alloy of the two metals. They are normally used where the rigid support of the copper is required in addition to the good qualities of lead. They are used as a backing for babbit, tin, silver, etc. in engine crankshaft bearings. They operate longer with less attention than many other metals.

6. Babbit bearings are an alloy of lead, tin, antimony, and often other metals. Being relatively soft, babbit allows some particles of foreign matter to become imbedded in it. This prevents wear on the shaft which would otherwise occur if the particles were not imbedded. These bearings are most commonly used as engine crankshaft and camshaft bearings.

7. Aluminum bearings have many applications where weight is a factor. Other uses are for high speed shafts and also some engine crankshaft bearings.

8. Sintered metal bearings (iron, brass, bronze, graphite, etc.) are commonly used on machines where minimum service is desired. They are constructed by subjecting fine particles of the material to tremendous pressure and heat. These bearings can absorb their lubricant and thus require very little service.

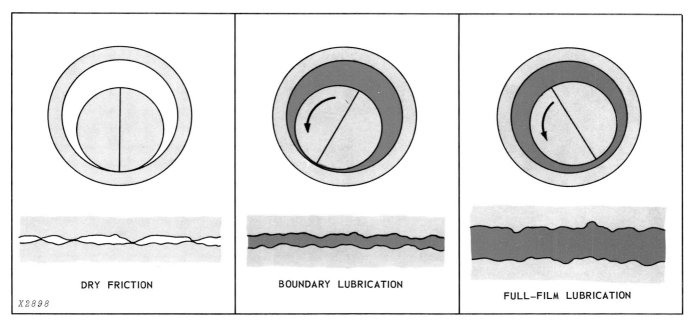

DRY FRICTION   BOUNDARY LUBRICATION   FULL-FILM LUBRICATION

X2898

Fig. 12 — Three Types Of Friction

9. Trimetal bearings are used for heavy duty engine crankshafts. They have a steel backing for load carrying capacity, a middle lining of a fatigue resistant copper-lead or aluminum alloy and a soft layer of lead-tin overplate for debris embedability and conformability.

10. Plastic bearings (phenolic, teflon, nylon, etc.) are becoming increasingly popular for bearings because of their resistance to corrosion, quiet operation, and conformability. They are also partly self-lubricating, which reduces the need for lubrication.

Laminated phenolic bearings consist of cotton fabric, asbestos, or other fillers, bonded with phenolic resin. They have good strength and shock resistance. Being relatively quiet, they are often used to make engine camshaft gears where noise may be a factor. However, precautions must be taken for adequate cooling as they are poor conductors of heat.

11. Rubber bearings are normally used as bearings for propeller and rudder shafts on boats and ships, hydraulic turbines, pumps, sand and gravel washers, and other machines which handle water or slurries. The resilience of rubber helps to reduce vibration and give quiet operation. It also allows operation with larger clearances and helps to compensate for misalignment.

LUBRICATING PLAIN BEARINGS

Lubricants are used to reduce friction, heat, and wear in bearings. The less friction and heat, the less wear.

Three states of friction or lubrication may exist between sliding surfaces (Fig. 12):

• **Dry friction**

• **Boundary lubrication**

• **Full-film lubrication**

DRY FRICTION is when no lubricant is present between the contacting surfaces, as in older crawler track pins and bushings, or disk harrows.

BOUNDARY LUBRICATION is when the lubricant is a thin film between the sliding surfaces. Only where movement is relatively slow and loads are light can this type of friction be acceptable.

FULL-FILM LUBRICATION separates the sliding surfaces completely with hydrodynamic pressure. This provides low friction and long wear life.

Notice, in Fig. 12, the three positions of the shaft as it changes from boundary lubrication to full-film lubrication. (The shaft load is downward). As the shaft begins rotating as shown in boundary lubrication, it tries to climb the side of the bearing. At moderate speed, the shaft returns and then begins to move to the opposite side as hydrodynamic pressure increases.

Full lubrication can only be achieved by a combination of the right lubricant, good bearing design, and at least a moderate rotary speed.

If the proper lubricant is not used, poor bearing performance will result.

For example, if the viscosity is too light, the fluid will not support the shaft load and metal-to-metal contact will occur.

If the lubricant is too heavy, the bearing will heat because of the friction created by the churning lubricant. This heat, if extremely high, will cause bearing failure.

How Lubricant Is Distributed To Bearings

The bearing must have these characteristics in order to perform well:

1. *It must have a means of distributing the lubricant.*

2. *Oil clearance must be correct to fit the shaft well.*

3. *The bearing surface must be compatible with the shaft, the lubricant, and the speed of rotation.*

OIL GROOVES

Oil grooves are necessary to insure that the bearing receives adequate distribution of lubricant. Several designs are shown below in Fig. 13.

BEARING CLEARANCE

Bearing clearance is very important. It depends upon load, speed, and type of lubricant.

If too little clearance is allowed, the load and the speed of shaft rotation will cause the bearing to overheat and fail.

Too much clearance will allow the lubricant to leak out and result in too little hydrostatic pressure, which will cause the surfaces to rub and wear.

BEARING SURFACE COMPATIBILITY

Fig. 14 on the next page shows the three states of lubrication. Each condition is determined by the viscosity of the lubricant, speed of shaft rotation, and smoothness of the mating surfaces.

Surfaces in Contact is caused by using a light-viscosity lubricant to support the shaft or by surfaces too rough to be separated by the recommended lubricant.

Surfaces Separated by Lubricant is the minimum lubrication requirement to prevent actual contact of mating surfaces. This condition exists with relatively slow speeds of rotation. However, if a lubricant with heavy viscosity is used here and shaft speeds are too fast, excessive heat will be generated by the friction of the churning lubricant.

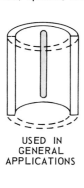 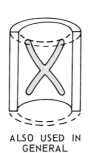 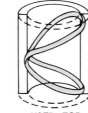 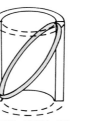 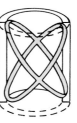 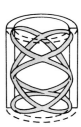

USED IN GENERAL APPLICATIONS | ALSO USED IN GENERAL APPLICATIONS | USED FOR FRACTIONAL HORSEPOWER MOTORS | USED WHEN OIL IS SUPPLIED NEAR END OF BEARING | USED FOR GREASE LUBRICATION | ALSO USED FOR GREASE LUBRICATION

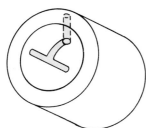 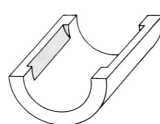 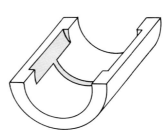 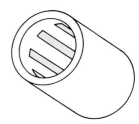

FEEDER GROOVE AND AXIAL DISTRIBUTING GROOVE USED FOR LARGE, SLOW SPEED BEARINGS. | CHAMFERED RECESS IN SPLIT BEARING DISTRIBUTES OIL ALONG SHAFT. USED IN ENGINES. | CIRCUMFERENTIAL GROOVE AND CHAMFERED RECESS IMPROVES OIL FILM COOLING. USED IN ENGINE BEARINGS. | USED FOR PULSATING LOADS WITH LIMITED SHAFT ROTATION.

Fig. 13 — Oil Groove Designs For Plain Bearings

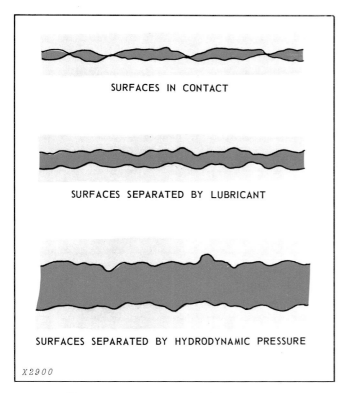

Fig. 14 — Lubricant Between Sliding Surfaces

Surfaces Separated by Hydrodynamic Pressure is the ideal state of lubrication. The surfaces of the mating parts, the speed of rotation, and the viscosity of the lubricant all affect the ability to achieve this condition. If the surface is too smooth and the shaft speed too slow, the hydrodynamic pressure will not be sufficient to lift the shaft; the surfaces may contact or be barely separated by the film of lubricant.

Methods Of Lubrication

Lubricants may be applied to plain bearings in many different ways. These methods vary from simple hand lubrication to complex pressure systems.

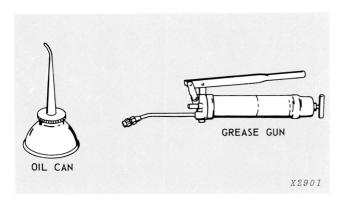

Fig. 15 — Two Types Of Hand Lubrication

Two methods of lubrication that are most familiar are the oil can and grease gun shown in Fig. 15. The oil can is used to apply lubricant directly on the parts or in an oil cup attached to the bearing. The grease gun injects grease through the fittings shown in Fig. 16.

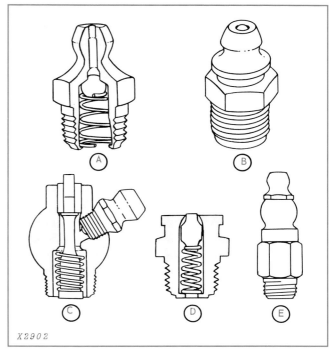

Fig. 16 — Types Of Grease Fittings

A. *Conventional Ball-Check Fitting*

B. *Pressure Relief Fitting*

C. *Pressure Shut-Off Fitting*

D. *Button-Head Fitting*

E. *Hydraulic Measuring Fitting*

Some of the most common types of grease fittings are shown in Fig. 16. Each fitting is designed for a particular requirement. Some fittings measure quantity or pressure and others relieve excessive pressure.

Another type of lubrication device is the central grease system which lubricates several, usually hard-to-reach, bearings with a few strokes of a plunger. Typical applications are on a hay baler or excavator boom.

One of the most common methods of oil lubrication is the oil bath splash system, such as shown in Fig. 17 on the next page. This method is used in simple engines, transmissions, differentials, and other components which can hold a large supply of lubricant. Rotation of gears splashes oil onto the vital bearing areas.

Fig. 17 — Oil Bath Splash System Used In Transmissions

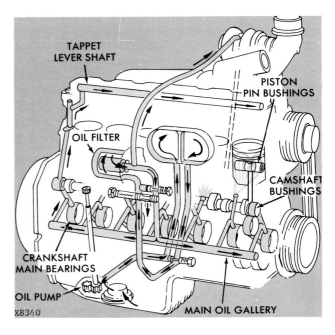

Fig. 18 — Pressure-Feed Oil System In A Typical Engine

Probably the best method of providing adequate lubrication to bearings is the pressure-feed system. In Fig. 18 you can see how this method is used to supply oil to critical bearings in an engine. Some complex machines also use a pressure system in their transmissions.

NOTE: For more information on lubricants, see the FOS manual on "Fuels, Lubricants, and Coolants."

REPAIR AND REPLACEMENT

Plain bearings are usually replaced when damaged or worn, seldom are they repairable.

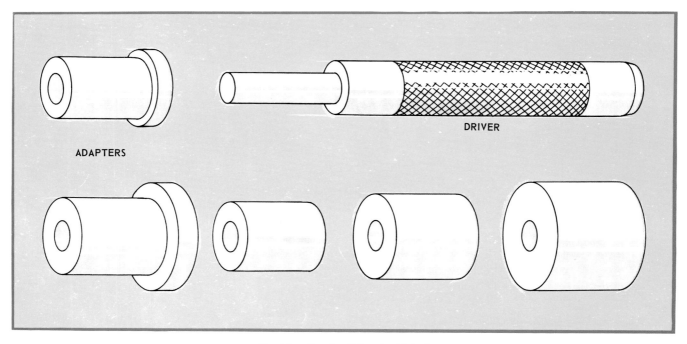

Fig. 19 — Bearing Driver And Adapter

Replacing Simple One-Piece Bearings

Bearing removal and installation is relatively simple with one-piece bearings. However, take precautions to prevent damage to the bore in which the bearing is installed. Use a length of cold rolled steel rod to drive the bearing in or out. Or, use a round nose cold chisel to cut a channel lengthwise through the bearing and drive the bearing out. However, use this method only if no other method is available.

A special driver, such as in Fig. 19, is more practical and less liable to damage the bore. Drivers can be purchased from your tool supplier with various adapters for different size bearings.

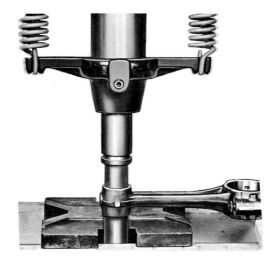

Fig. 21 — Installation Of Bearing With A Press

When possible, use a press as shown in Fig. 21 to install bearings. This assures that the bearing is pressed into its bore accurately. Care must be exercised in aligning lubrication holes or passages in the bearing with those in the component.

In some cases it is necessary to ream a bearing to size where a critical fit is required. A pilot and reamer are used to size the bearing bore as shown below in Fig. 22. If a pilot is not used, the reamer may not start correctly. Always be sure that the reamer or hone is inserted straight into the bearing bore. If not, the bore will not be square with the shaft. There will not be a full bearing and rapid wear will occur.

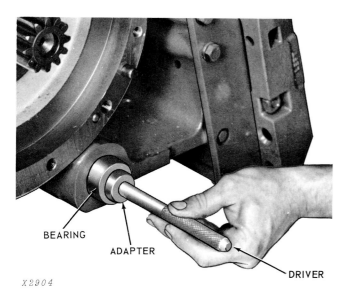

X2904

Fig. 20 — Bearing Installation With A Driver And Adapters

Use extreme care when installing bearings (Fig. 20). The bore must be clean and free from burrs. Drive the bushing *straight* into the bore. If it is "cocked" when driven in, it will be distorted and damaged.

Replacing Split Bearings

Split bearings are most often used in engine crankshafts; they are also used in many other applications. However, the same basic principles apply when replacing all these types of bearings.

NOTE: For more information about engine bearing service, refer to the FOS manual on "Engines."

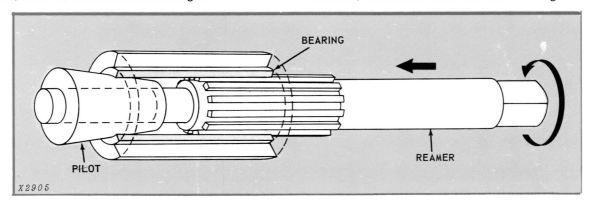

X2905

Fig. 22 — Reaming Bushing To Size

Measuring Bearing Clearance

Before new bearings can be installed, you must determine the condition of the shaft. If the shaft surface is too worn or rough, it may have to be reground or replaced. Check for journals that are tapered or out-of-round. If the crankshaft needs regrinding, see Fundamentals of Service—Engines (Chap. 2) for the correct procedure.

Fig. 23 — Measuring Shaft Wear

If the shaft journals appear to be in good condition, measure them in several places around the shaft to determine the amount of wear (Fig. 23)., Refer to your machine Technical Manual for wear limits.

Fig. 24 — Measuring Bearing Clearance

Install new bearings and tighten bearing caps to specified torque. Using an inside micrometer, measure the inside diameter of the bearing (Fig. 24). Compare the reading with the outside diameter of the shaft. Determine the clearance by subtracting the shaft diameter from the bearing inside diameter. Refer to specifications to determine the correct clearance.

Under-sized bearings may have to be installed if too much wear has occurred on the crankshaft journal.

Measuring Bearing Clearance Using Plastigage

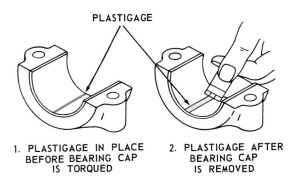

1. PLASTIGAGE IN PLACE BEFORE BEARING CAP IS TORQUED   2. PLASTIGAGE AFTER BEARING CAP IS REMOVED

Fig. 25 — Using Plastigage To Measure Bearing Clearance

Another method of measuring bearing clearance is by using "Plastigage." This is a plastic thread which "crushes" to the exact clearance when the bearing cap is fully tightened. While this method will give the bearing clearance, it will not tell you whether the wear is on the bearing or on the shaft.

Installing Split Bearings

Extreme care must be taken when installing split bearings. The bearing must be round and true, since there must be perfect contact between the inside of the bearing seat and the outside of the bearing. This is one factor that is often overlooked when these bearings are installed. If the contact isn't perfect, the heat will not flow from the bearing to its support, and the bearing may melt. This is one reason why shims must not be used between the bearing and its seat to correct wear. When installing the bearing, be sure that oil passages are aligned and that the locking devices are in correct relationship (Fig. 26 on next page).

The need for heat dissipation is one reason for bearing *crush*. Crush means that each half of the bearing inserts extends one or two thousandths of an inch (0.025 or 0.050 mm) beyond the bearing seat bore as

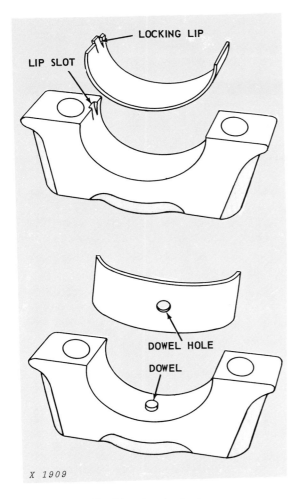

Fig. 26 — Bearing Locks

shown in Fig. 27. Then when the cap screws of the bearing cap are tightened to the specified torque, the insert is forced to seat solidly in the bearing seat.

Another reason for crush is to insure that the bearing remains round. If not tightly held on the edge, it might distort as illustrated in exaggerated form in Fig. 28 on the next page.

Still another reason for crush is to avoid any possible movement of the bearing in the seats. If the bearing becomes loose, it could oscillate in its seat and wear on the outside. This could interfere with oil flow and heat transfer. Any dirt between the insert and its seat would have the same effect.

On the other hand, the amount of crush must not be excessive; only 0.001 or 0.002 inch (0.025 or 0.050 mm). If it is more than 0.002 inch (0.050 mm), the insert will be distorted as illustrated in Fig. 28, when the cap is tightened. The bearing will buckle and result in increased friction and heat; leading to bearing failure.

Also inspect joint surfaces of cap, connecting rod, and block for excessive wear, nicks or other irregularities. Use correct procedures for tightening cap screws. These items all have an effect on proper bearing crush and joint fit.

Split bearings are extremely accurate as they come from the manufacturer and must be handled with care. Always use the precise size required and install without alteration or fitting.

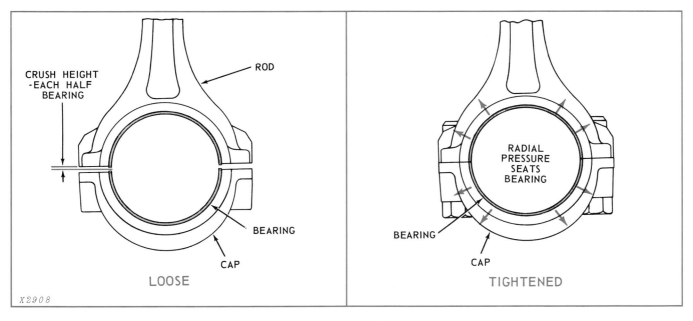

Fig. 27 — Bearing "Crush"

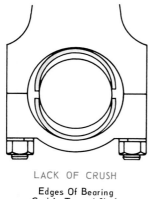

LACK OF CRUSH

Edges Of Bearing
Curl In Toward Shaft

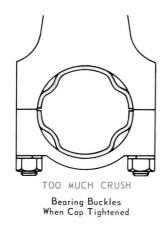

TOO MUCH CRUSH

Bearing Buckles
When Cap Tightened

X2909

Fig. 28 — Effects Of Improper Crush

DIAGNOSING PLAIN BEARING FAILURES

When replacing damaged bearings, it is vital to determine the *cause* of failure and correct it. Otherwise, failures may occur again from the same cause.

Most bearing failures can be linked to *specific* causes. One bearing manufacturer found that causes of failures ranked as follows:

1. **Lack of Lubrication**
2. **Improper Assembly**
3. **Corrosion**
4. **Dirt**
5. **Misalignment**
6. **Tapered Bearing Journals**
7. **Overloading**

When worn or damaged bearings are removed, examine them using this section as a guide.

Lack of Lubrication

R 6317

Fig. 29 — Oil Starvation Caused This Damage

Oil starvation was the cause of damage to the bearing shown in Fig. 29. Lack of oil can occur immediately after overhaul. This is when priming of the lubricating system is vital to assure initial lubrication.

After break-in, other things can happen. Both local and general oil starvation can result from external leaks and mechanical supply failures. Blocked oil suction screen, oil pump failure, oil passages plugged or leaking, failed pressure relief valve springs, or badly worn bearings can stop the circulation of lubricating oil.

Also, in the case of engines, the oil supply may become diluted by seepage of fuel into the crankcase from a defective fuel pump. This will reduce the oil's film strength and the increased friction will score the bearings.

Improper Assembly

A mislocated oil hole will also cut off the oil supply to a bearing, causing rapid failure. Always check to be sure that the oil hole in the bearing is in line with the oil supply hole.

R 6318

Fig. 30 — Corrosion From Acid Formation In Oil

Corrosion

Corrosion from acid formation in the oil is seen by a finely pitted surface and large areas of deterioration (Fig. 30).

Corrosion occurs when oil temperature goes above 300° F (150° C) and when excessive blow-by occurs in engines. It is further aggravated by condensation. In some cases, incorrect lubricant will also cause corrosion.

Prevent corrosion in engines by changing oil at correct intervals and by selecting oil of the proper quality and classification for the machine and type of service. Follow the manufacturer's recommendations.

Dirt

R 6319

Fig. 31 — Damage From Dirt Embedded In Bearing

Large *dirt* particles can embed in the soft bearing material (Fig. 31). This causes wear and decreases the life of both the bearing and its journal.

Prevent this by cleaning the seating area of the bearing thoroughly during bearing installation and by proper maintenance of both air and oil filters, if used.

BEARING BACK

BEARING FACE

R 6320

Fig. 32 — Particle Of Dirt Left From Improper Cleaning

A particle of dirt was left (Fig. 32, arrow) during bearing installation. It remained behind this bearing on its cap to push the bearing inward at one point, thus

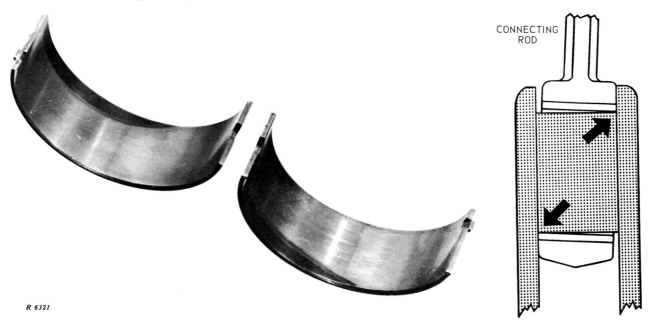

R 6321

CONNECTING ROD

Fig. 33 — Excessive Wear Caused By Misalignment

Fig. 34 — Wear On One Edge Of Bearing Caused By Tapered Journals

increasing localized pressure and heat. This damage started because of improper cleaning of a bearing. Dirt was allowed to remain and create an area of excessive heat that led to this damage.

Misalignment

Misalignment can cause concentrated wear on the bearings (Fig. 33). This angular loading will cause the excessive wear shown on one edge of the upper bearing and the opposite edge of the lower bearing. When this wear pattern is found, check the alignment of the connecting rod and bearings to the shaft.

Tapered Bearing Journals

Tapered journals allow areas of excessive clearance between the journal and bearing; this distributes more wear on one edge of the bearing (Fig. 34). This wear is increased by the force on the bearing carrying the greater load.

Overloading

Fig. 35 — Bearing Fatigue Caused By Overloading And Heat

Overheating from overloads causes a metal fatigue which breaks away and voids the surface of the bearing (Fig. 35).

TROUBLE SHOOTING OF PLAIN BEARINGS

We have already discussed bearing failures by examining the bearing. Now let's learn about symptoms that indicate possible bearing failure. Some of the obvious signs are a sudden or sustained rise in temperature, excessive noise, rough operation, and excessive bearing and shaft clearance.

These Trouble Shooting charts are a general guide to help determine the possible cause and remedy of bearing failures.

Additional trouble shooting information of plain bearings can be found in the FOS manual, Identification of Parts Failures, chapter 2.

Keep in mind that many of these symptoms in the following chart can be present at once—and several causes can occur at the same time.

Following the manufacturer's break-in procedures will greatly reduce bearing failures. To detect impending bearing failures, an oil analysis on a regular basis can be a valuable preventive maintenance approach.

Possible Causes	Symptoms			Remedy
	Excessive Temperatures	Excessive Noise and Rough Operation	Excessive Wear	
Shaft and bearing misaligned.	X	X	X	Realign.
Bearing caps too tight.	X		X	Retighten to specification or replace bearings with correct size.
Bearing overloaded.	X	X	X	Reduce load or speed.
Fatigue failure.	X	X		Replace bearing.
Faulty oil grooving.	X	X	X	Replace bearing.
Bearing surfaces too rough.	X	X		Complete break-in period or use suitable bearing.
Bearing clearances too small.	X	X		Replace with correct size bearing.
Unsuitable bearing materials because of poor fatigue resistance, conformability, embeddability, corrosion resistance, or thermal conductivity.	X	X		Replace with suitable bearing.
Faulty bearing design or construction.	X	X	X	Replace bearing.
Abrasive or corrosive contaminants in lubricant.	X	X	X	Replace lubricant.
Lubricant too light for imposed loads.	X	X	X	Use heavier lubricant.
Lubricant too heavy for bearing clearance.	X			Use lighter lubricant.
Lubricant too heavy for speed of rotation.	X			Use lighter lubricant.
Lubricant unsuitable for application.	X	X	X	Use recommended lubricant.
Oil passages blocked.	X	X	X	Clear oil passages.
Excessive vibration.		X	X	Eliminate cause of vibration.
Bearing cap loose.		X	X	Tighten to specification.
Excessive bearing clearance.		X		Replace with correct bearing.
Excessive leakage.		X	X	Replace seals.

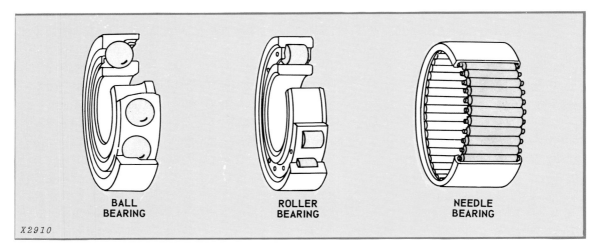

Fig. 36 — Three Basic Types Of Anti-Friction Bearings

ANTIFRICTION BEARINGS

There are three basic types of antifriction bearings (Fig. 36).

• **Ball bearings**

• **Roller bearings**

• **Needle bearings**

Most antifriction bearings are made of hardened steel. However, some bearings are made of other materials, such as plastics.

BASIC PARTS

Most antifriction bearings have similar parts:

1. *Two hardened-steel rings called* **races** *(Fig. 37).*

2. *Hardened-steel* **balls, rollers** *or* **needles** *which roll between the races.*

3. *Optional* **separators** *or cages (made of steel, bronze or plastic) which space the rolling elements around the races.*

On some bearings the outer or inner race is omitted, and the rolling elements are in direct contact with the shaft or other mounting (as in many needle bearing applications).

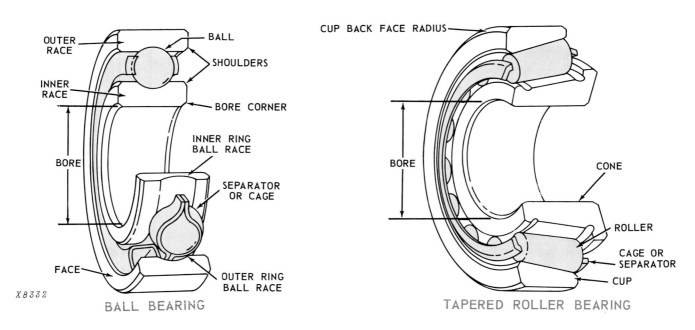

Fig. 37 — Basic Parts of Antifriction Bearings

When two races are used, one race is normally pressed or fixed on a shaft or in a bore, while the other race is free to turn with the rolling elements. This is part of the "anti-friction" feature of these bearings.

We have already discussed the purposes, general types, advantages and disadvantages of all bearings in the introduction to this chapter. Before examining specific types of bearings, let's take a look at bearing capacities.

BEARING CAPACITIES

The load-carrying capacity of a bearing is determined by three conditions:

1. **Size of bearing.**
2. **Number of rolling elements.**
3. **Type of races.**

Some other factors, such as surface finish hardness, may also have an effect on bearing capacities.
Let's look at each factor.

Size of Bearing

The *size of a bearing* is the primary factor which determines load-carrying capacity: A larger bearing will support more load than a smaller one of the same design.

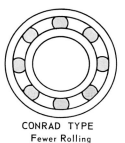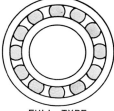

CONRAD TYPE
Fewer Rolling Elements

FULL TYPE
More Rolling Elements

X2912

Fig. 38 — Number Of Rolling Elements

Number of Rolling Elements

The *number of rolling elements* is also important. In Fig. 38 you can see that the Conrad ball bearing has fewer balls than the full bearing. The full bearing can support a greater radial load than the Conrad bearing. However, the Conrad bearing can withstand more thrust load than the full bearing. This is due to the race design, as you will see later. Often, the load-carrying capacity of a bearing is doubled by using a double row of balls or rollers.

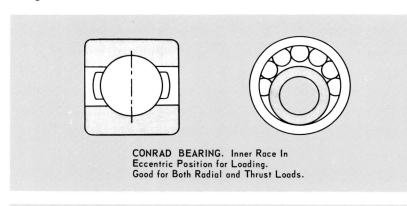

CONRAD BEARING. Inner Race In Eccentric Position for Loading.
Good for Both Radial and Thrust Loads.

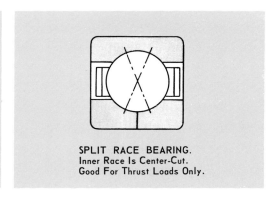

SPLIT RACE BEARING.
Inner Race Is Center-Cut.
Good For Thrust Loads Only.

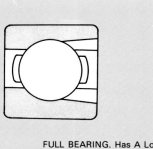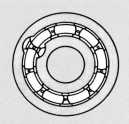

FULL BEARING. Has A Loading Slot.
Holds More Balls, Only For Radial Loads.

X2913

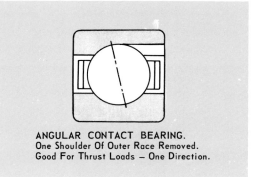

ANGULAR CONTACT BEARING.
One Shoulder Of Outer Race Removed.
Good For Thrust Loads — One Direction.

Fig. 39 — Types Of Ball Bearing Races

Types of Bearing Races

The *type of races* in the bearing controls the load-carrying capacity, especially in ball bearings. Most roller and needle bearings have only two variations of races: split and continuous. Because of the load distribution of their rolling elements, these bearings are not affected greatly by the split race.

Ball bearings, on the other hand, have a variety of races. Basically, ball bearings have four race designs.

1. **Conrad Bearing**
2. **Full Bearing**
3. **Split Race Bearing**
4. **Angular Contact Bearing**

It is important that a replacement bearing have the same load-carrying capacity as the original.

If, for example, you replace a Conrad bearing with a full bearing (Fig. 39) you will risk premature bearing failure.

The reason for this is that the *Conrad* bearing has a deep groove with continuous race shoulders. This gives it both good radial and thrust capacities. As you can see, the Conrad bearing is loaded with the inner race displaced. After loading, the inner race is moved back to center, the balls are evenly positioned, and the separators are installed.

The *full* bearing has a loading slot which allows more balls to be installed than the Conrad bearing. Because of this loading slot, the full bearing cannot take the thrust loads which the Conrad bearing can. For this reason, the full bearing is recommended where only radial loads are present.

Split race bearings (either inner or outer race) are essentially angular contact bearings that can take high thrust loads in either direction. This is because of their high shoulders and the ability to load a full complement of balls because of the removable split races. These bearings cannot accept high radial loads.

The *angular contact* bearing (Fig. 39) can accept both radial and thrust loads, but thrust loads in one direction only. However, it cannot withstand as much radial load as the Conrad bearing.

TYPES OF BALL BEARINGS

Ball bearings are designed to carry different loads. Sizes range from fractions of an inch to several inches in diameter.

Since a given bearing is used for a specific job, such as a radial load, it must not be subjected to thrust loads or it will fail prematurely.

Another thing to keep in mind is that ball bearings are precision made and finely finished. They must be handled with care and kept clean.

Notice in Fig. 40 that the center lines of force and the areas of contact are shown for each bearing.

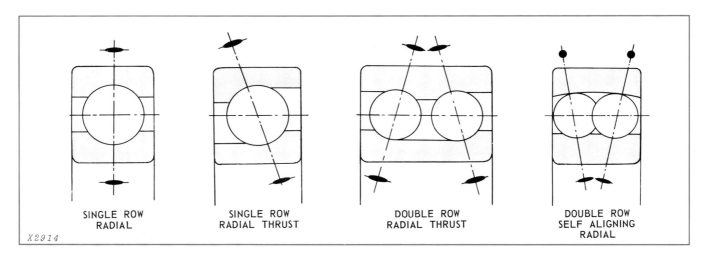

SINGLE ROW
RADIAL

SINGLE ROW
RADIAL THRUST

DOUBLE ROW
RADIAL THRUST

DOUBLE ROW
SELF ALIGNING
RADIAL

X2914

Fig. 40 — Types Of Ball Bearings

The four basic designs of ball bearings are:

1. **Radial Load**
2. **Radial and Thrust Load**
3. **Self Aligning, Radial Load**
4. **Thrust Load**

Radial load ball bearings are available with either single row or double row (Fig. 40). Some have shields that prevent foreign matter from entering the bearing and seals that retain the lubricant and exclude dust. Though these bearings primarily support radial loads, they will accept some thrust loads, providing the load is not too great. The grooves of radial bearings are shallow with little shoulder to support thrust loads.

Radial and thrust load ball bearings are also available as singles or doubles (used in pairs or double rows), and they also have optional shields and seals available. As singles, these bearings will support radial loads and thrust loads in one direction against the higher shoulder of the outer race (Fig. 40). When singles are combined into duplexes, they can take thrust loads in either direction, as can the double row design.

Self aligning radial load ball bearings can be either internally self aligning or externally self aligning. They can accept either *initial* misalignment (by having a spherical outer race) or they can accept *running* misalignment (by having a spherical roller path). The first type is usually a sealed bearing. Fig. 40 shows the internally self-aligning bearing.

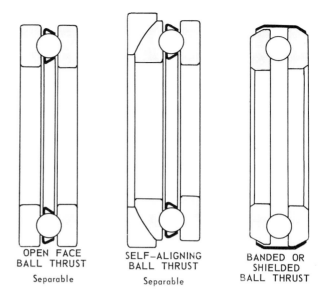

OPEN FACE BALL THRUST
Separable

SELF-ALIGNING BALL THRUST
Separable

BANDED OR SHIELDED BALL THRUST

Fig. 41 — Thrust-Load Ball Bearings

Thrust load ball bearings (Fig. 41) are usually single row and may have a shield or seal. These bearings will accept full thrust loads and some light radial loads. They are usually used in conjunction with radial load bearings. In Fig. 41 you can see that three varieties are available. The open face and self-aligning designs are separable, but the shielded is not.

TYPES OF ROLLER BEARINGS

Where great loads are present, roller bearings are used instead of ball bearings.

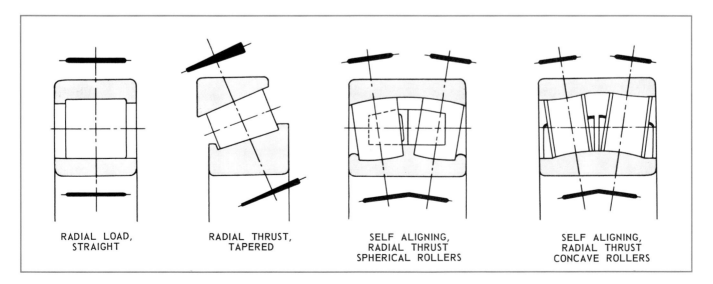

RADIAL LOAD, STRAIGHT

RADIAL THRUST, TAPERED

SELF ALIGNING, RADIAL THRUST SPHERICAL ROLLERS

SELF ALIGNING, RADIAL THRUST CONCAVE ROLLERS

Fig. 42 — Basic Types Of Roller Bearings

Roller bearings are also designed for loads and applications similar to those of ball bearings. As with ball bearings, roller bearings are precision made and must be handled with care and kept clean.

Notice in Fig. 42 that the center lines of force and the areas of contact are shown for each bearing.

The five basic types of roller bearings are:

1. **Radial load, straight roller**
2. **Radial and thrust load, tapered roller**
3. **Self aligning, radial and thrust load, spherical roller**
4. **Self aligning, radial and thrust load, concave roller**
5. **Thrust load**

Radial load, straight roller bearings have high radial capacity, but they cannot accept much misalignment or thrust loads (Fig. 42). They are also available with two or more rows of rollers. Some are nonseparable while others have separable inner or outer rings. These bearings are normally used in applications where high speeds and heavy loads are imposed.

Radial and thrust load, tapered roller bearings will accept combined heavy radial and thrust loads (Fig. 42). They are normally used in pairs or double row units for moderate and low speeds. These are the most popular types of roller bearing. They are adjustable through their separable outer cups or through their cones. Some double row units are preadjusted.

Self aligning, radial and thrust load, spherical roller bearings (Fig. 42) have great load carrying capacity due to the number, size, and shape of the rollers and the spherical aligning surface. Because these bearings are self aligning, some misalignment will not affect their performance. These bearings can accept more thrust loads than straight rollers, but less thrust than tapered rollers. The single row, radial bearing cannot accept much thrust load. The single row, radial and thrust bearings can only accept thrust loads in one direction. The double row bearing, however, can withstand thrust loads in either direction. Obviously this bearing offers more versatility than any other roller bearing which is important when a bearing must function under a variety of conditions.

Self aligning, radial and thrust load, concave roller bearings (Fig. 42) have the same load carrying characteristics as the spherical roller bearings. The basic differences are that the rollers are concave rather than convex, and the spherical self-aligning surface is on the *inner* ring rather than on the outer.

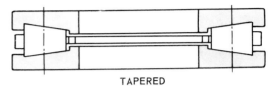

TAPERED

SPHERICAL SELF-ALIGNING

Fig. 43 — Trust Load Roller Bearings

Thrust load roller bearings (Fig. 43) can carry heavy thrust loads, but they cannot withstand any radial loads. The *tapered* roller bearing (top) cannot accept misalignment like the *spherical* self-aligning roller bearing (bottom) can. Normally, thrust load roller bearings are used in conjunction with radial load bearings.

TYPES OF NEEDLE BEARINGS

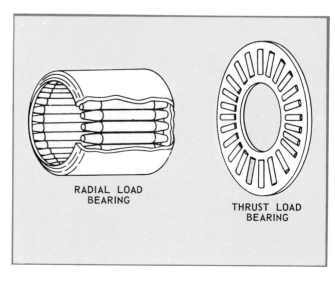

RADIAL LOAD
BEARING

THRUST LOAD
BEARING

Fig. 44 — Basic Types Of Needle Bearings

Needle bearings have very small diameter rollers or needles, but high load carrying capacity (Fig. 44). They are most often used where space is limited, but they do not perform at high speeds as well as roller or ball bearings. Also, they will not tolerate much misalignment.

There are two main types of needle bearings:

• **Radial load bearings**

• **Thrust load bearings**

The radial load needle bearing (Fig. 44) may or may not have an inner sleeve or race. The *thrust load bearing* can be used with or without races. If used without races, the mating surfaces must be ground square and hardened. The needles may be either caged, retained by grease, or loose for hand assembly.

MOUNTINGS FOR ANTIFRICTION BEARINGS

Antifriction bearings must be mounted so that their shape is not distorted, the rolling elements are not bound, and the inner and outer races are aligned (not

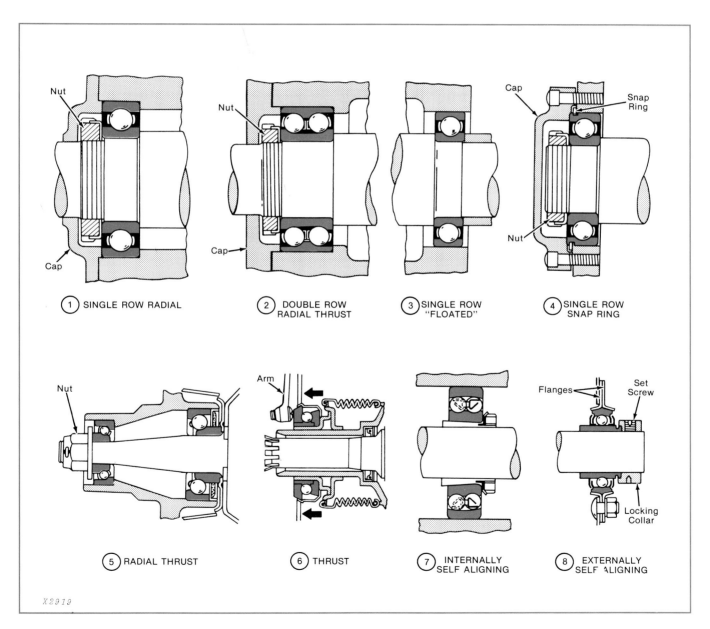

1 SINGLE ROW RADIAL

2 DOUBLE ROW RADIAL THRUST

3 SINGLE ROW "FLOATED"

4 SINGLE ROW SNAP RING

5 RADIAL THRUST

6 THRUST

7 INTERNALLY SELF ALIGNING

8 EXTERNALLY SELF ALIGNING

X2919

Fig. 45 — Typical Ball Bearing Mountings

necessary with self-aligning bearings). In addition, when two or more bearings are mounted on a common shaft, the axis of each bearing must be aligned with the other.

Many antifriction bearings have seals to retain lubricant and exclude dirt. Antifriction bearings may also be lifetime lubricated and sealed on both sides. See the section on "Seals" later in this manual.

Antifriction bearings are usually mounted with one race a push fit and the other a press fit. A push fit means that the bearing race can be slid into place by hand. A press fit requires a press or a driver to install the race.

Remember that normally the press fit race is pressed onto or into the *rotating* part, and the push fit race is pushed onto or into the *stationary* part. However, this rule of thumb is not true in every situation. Where large bearings encounter tremendous loads and speeds, *both* races may be *pressed* into place. These bearings must have internal clearances built in.

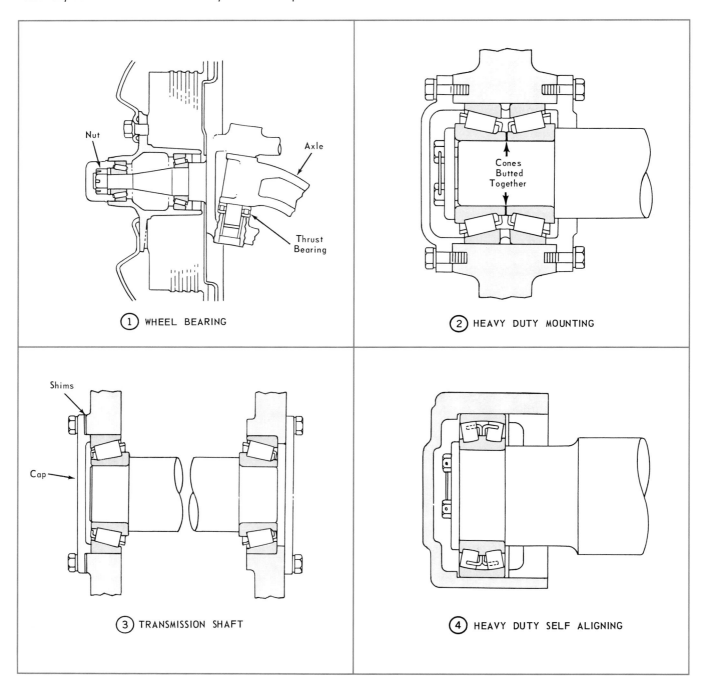

Fig. 46 — Typical Roller Bearing Mountings

Ball Bearing Mountings

A typical method of mounting a single row, radial bearing is shown at upper left in Fig. 45. The bearing is held in position against the shoulders by the nut and cap.

The double row, radial thrust bearing (2) is mounted similarly (Fig. 45), but great importance is placed on preventing movement due to the thrust loads that may be imposed upon it.

Often, the other end of the shaft has a single or double row, "floated" bearing (as shown in 3, Fig. 45). This bearing is allowed to shift in the bore rather than accept thrust loads; hence the description "floated."

When the bore must be straight through (without shoulders), a snap ring is often used (as shown in 4, Fig. 45) to hold the bearing in position.

Radial thrust bearings (5) are usually mounted in wheels as shown. The outer race is pressed into the bore and the inner race is a push fit. Adjustment is made with the nut.

Thrust bearings (6) are mounted for clutch disengagement as shown. The arm disengages the clutch because of pressure applied through the bearing.

Internally self-aligning bearings (7) may be mounted as a "floating" bearing, as shown, or they may be retained by a nut and cap.

The externally, self-aligning bearing (8), is often mounted with flanged retainers as shown. Similar bearings may be mounted in a spherically machined housing. A locking collar is used to secure the bearing. (The collar must always be tightened in the direction of shaft rotation, then locked with a set screw.)

Roller Bearing Mountings

One of the most common methods of mounting tapered roller bearings is shown in Fig. 46, No. 1. Typical wheel bearings are shown. The outer race or cup is pressed into the housing and the inner race or cone is pushed onto the shaft. Adjustment is made by turning the nut.

Also shown in Fig. 46, (1) is a thrust bearing. In this case the thrust bearing is used to support the load of the axle.

In heavy duty applications, double roller bearings are used (2 in Fig. 46). Here the outer race or cup is one piece and the cones are two singles.

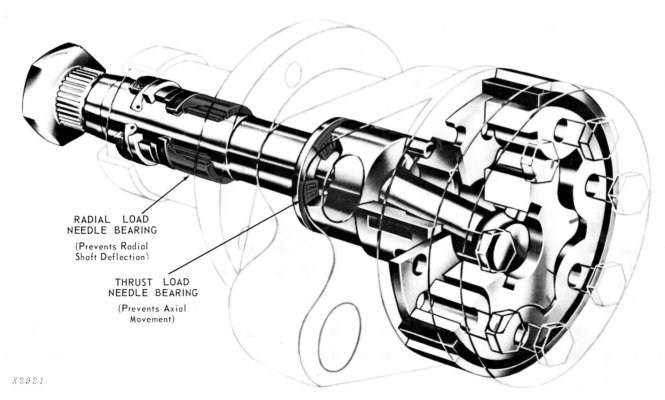

RADIAL LOAD
NEEDLE BEARING

(Prevents Radial
Shaft Deflection)

THRUST LOAD
NEEDLE BEARING

(Prevents Axial
Movement)

X2921

Fig. 47 — Typical Needle Bearing Mountings

This bearing is preadjusted by butting the cones together (spacers can also be used). Another common mounting method is used in transmissions (3 in Fig. 46). The transmission shaft is supported at both ends by tapered roller bearings as shown. Adjustment is made by using shims, as necessary, under the cap.

One type of mounting for heavy duty self aligning bearings is shown in Fig. 46, No. 4. This bearing and mounting is used where heavy loads are encountered and alignment is a problem.

Needle Bearing Mountings

Two types of needle bearings are shown mounted in a hydraulic steering pump in Fig. 47.

The radial load bearing accepts radial loads and prevents shaft deflection, while the thrust load bearing prevents axial movement of the shaft.

Another common use of radial load needle bearings is in universal joints where space and weight is at a premium and loads are high.

In some needle bearings, the needles must be assembled by hand during installation. The needles may be packed in heavy grease, retained by rubber bands, or by an adhesive strip until the bearing is in place. Be careful not to lose any needles.

TIPS FOR ANTIFRICTION BEARING MAINTENANCE

Do

1. Do work with clean approved tools, in clean surroundings.

2. Do clean outside of housings before exposing bearings.

3. Do handle bearings with clean, dry hands, or better, use clean canvas gloves.

4. Do work on a clean metal or metal covered bench.

5. Do treat a used bearing as carefully as a new one, until the used one is proven to be defective.

6. Do use clean solvents and flushing oils.

7. Do lay bearings out on a clean surface.

8. Do protect disassembled bearings from dirt and moisture.

9. Do wipe bearings, if necessary, **only** with clean, lint free towels.

10. Do keep bearings wrapped in oil-proof paper when not in use.

11. Do thoroughly clean the inside of housings before installing bearings.

12. Do install new bearings as they come from the package, without washing, if they are received in a sealed container.

13. Do keep lubricants clean when applying them, and cover the containers when not in use.

Don't

1. Don't use wooden mallets.

2. Don't handle bearings with dirty or moist hands.

3. Don't spin uncleaned or dry bearings.

4. Don't spin **any** bearing with compressed air.

5. Don't use the same container for both cleaning and final rinse of used bearings.

6. Don't use cotton waste or dirty rags to wipe bearings.

7. Don't scratch or nick bearing surfaces.

8. Don't crack or chip the bearing by hammering it into place.

9. Don't use incorrect type or amount of lubricant.

 CAUTION: Don't use gasoline for cleaning bearings. Gasoline is a highly flammable liquid. Gasoline containing tetraethyl lead is also a health hazard.

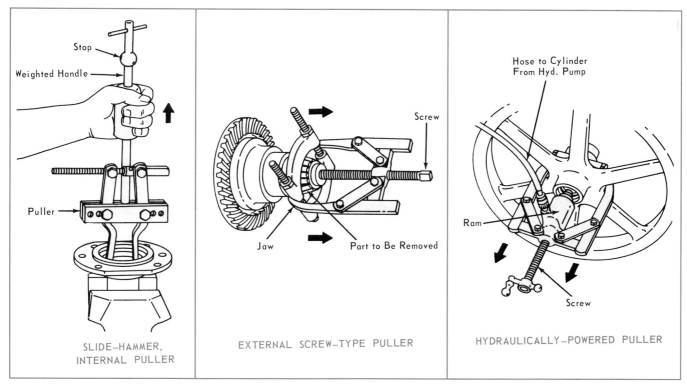

Fig. 48 — Typical Bearing Pullers

MAINTENANCE OF ANTIFRICTION BEARINGS

The most important factor for long bearing life is good bearing maintenance. This includes proper lubrication, repair, diagnosis of failures, installation, and adjustment. Some of the key "do's" and "don'ts" are shown on page 26.

Let's examine bearing maintenance, starting with bearing removal.

 CAUTION: Refer to pages 85–87 for important safety information.

Removing Bearings

Before attempting to remove any bearing, clean the bearing housing and shaft. Consult the Technical Manual or Service Manual for removal procedures. If no procedures are given, study the bearing mounting carefully to determine how it must be removed. The maintenance tips on the preceeding page are important when working with bearings.

Bearings can be removed three ways:

1. *Using either mechanical or hydraulic pullers*

2. *Using either mechanical or hydraulic presses*

3. *Using a hammer and proper driver*

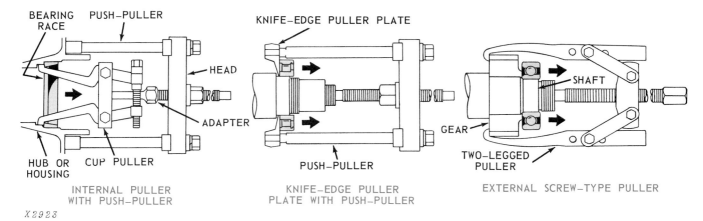

X2923

Fig. 49 — Correct Puller Applications

REMOVING BEARINGS USING PULLERS

Three types of bearing pullers are shown in Fig. 48. The *slide-hammer puller* is shown in the first drawing. By sliding the weighted handle against the stop, the bearing is forced out of the housing.

Another type of puller is the *screw-type puller* which pulls the bearing from the shaft by the force of the screw turning.

The *hydraulically-powered puller* is the most powerful. Some models of hydraulic pullers are capable of several tons of force.

Pullers may be for either internal or external applications, depending on the need. Some pullers can be used for both jobs because they have reversible jaws with both external and internal ends.

Fig. 49 illustrates some common puller applications.

The first puller shown is an *internal* type used with a push-puller. The push-puller may also be used for installation, as we will see later. The legs of the push puller support the head while the screw is turned, pulling the bearing cup from the housing.

A *knife-edge puller* plate is available to remove bearings where other pullers would damage the bearing. This plate is shown at center in Fig. 49 removing a bearing from a shouldered shaft. If the regular external-type puller were used to remove this bearing, the force exerted on the outer race would damage the bearing. Notice that the knife edge plate is used with a push-puller; the external screw-type puller can also be used with the knife edge plate.

The *external screw-type puller* can be used to remove bearings as shown at right in Fig. 49 provided there is some solid object which allows the jaws of the puller to force the bearing off.

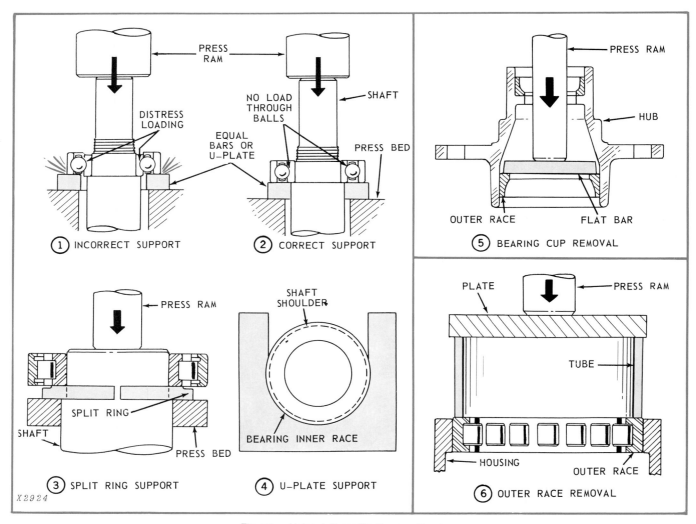

Fig. 50 — Using A Press To Remove Bearings

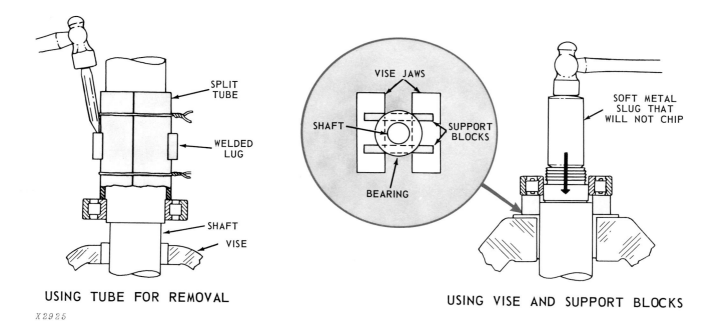

SPLIT TUBE

WELDED LUG

SHAFT

VISE

USING TUBE FOR REMOVAL

X2925

VISE JAWS

SHAFT

SUPPORT BLOCKS

BEARING

SOFT METAL SLUG THAT WILL NOT CHIP

USING VISE AND SUPPORT BLOCKS

Fig. 51 — Correct Methods Of Removing Bearings With A Hammer

Many variations of these basic pullers are available in several sizes and capacities. Pullers are one of the best tools to use to remove bearings easily and safely and to avoid damage to the bearing or its mating part.

REMOVING BEARINGS USING A PRESS

Another excellent tool for removing bearings correctly is the press. Either a mechanical or hydraulic press can make bearing removal easy. *Avoid excessive force.*

When removing a bearing from a shaft, be sure that the inner race is supported so that the outer race is not distress loaded. If force is transmitted through the outer race, damage may result to the bearing.

In Fig. 50 various methods of removing bearings with a press are shown.

In example, No. 1, *incorrect* support of the inner race is shown. This forces the balls against the shoulders of the bearing races, which could result in chipping or cracking the races or balls. Example No. 2 shows the *correct* support of the inner race. Split rings (3) or U-plates (4) are also available to provide correct support.

Bearing cups (5) may be pressed from the housing using a flat bar which transmits ram pressure. This can only be used when the housing is open from the

opposite side and the press has enough clearance to accept the housing.

To press the outer race from a housing (6), use a tube that is slightly smaller in diameter than the outer race. Place a plate over the tube and press the bearing from the housing.

REMOVING BEARINGS USING HAMMERS AND DRIVERS

If a bearing puller or a press is not available, the bearing can be removed with a hammer and a driver. In Fig. 51 you can see two methods of removing bearings from a shaft.

 CAUTION: Always wear safety glasses when driving a bearing with a hammer.

The illustration at left in Fig. 51, shows the use of a tube placed over the shaft to drive the bearing from the shaft. If the shaft has obstructions, the tube may be split and fastened to the shaft as shown. The welded lugs on the side of the tube provide a surface to strike with the hammer and punch.

The other method, shown at right in Fig. 51, is similar to what we described for presses. Use support blocks, split rings, or a U-plate to support the inner race of the bearing. Then drive the shaft from the bearing with a soft slug or driver.

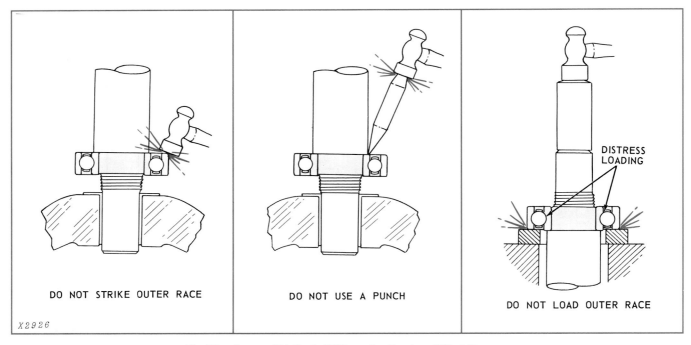

X2926

Fig. 52 — Incorrect Methods Of Removing Bearings With A Hammer

| DO NOT STRIKE OUTER RACE | DO NOT USE A PUNCH | DISTRESS LOADING — DO NOT LOAD OUTER RACE |

The three incorrect methods of removing bearings shown in Fig. 52 will probably result in bearing damage:

1. *Do not strike the outer race. The races or balls will chip or crack.*

2. *Do not use a punch as shown; it may slip and damage the balls or races.*

3. *Do not load the outer race. Be sure that the supports are under the inner race or damage will result.*

Cleaning And Inspecting Bearings

After removal, bearings must be cleaned before inspection. If the bearing is sealed and does not have removable shields, it cannot be cleaned. These bearings can be checked only for external damage or for smooth turning. Bearings with a shield on one side only can be cleaned and inspected just as those without shields.

 CAUTION: Never use gasoline for cleaning bearings. It is highly flammable. Gasoline containing tetraethyl lead is also a health hazard.

Always follow the manufacturer's recommendations for cleaning bearings. Use only petroleum solvents that are recommended for washing. Other solvents may be used, such as kerosene, naptha, etc., but they may not be as safe nor clean as well as solvents formulated for washing purposes. Most all solvents are very flammable and precautions must be taken to prevent fire.

 CAUTION: Many solvents are poisonous when inhaled or absorbed into the skin. Wear protective gloves and a chemical cartridge respirator.

Fig. 53 — Cleaning Bearings In A Container Of Safe Solvent

Do not try to clean bearings in a container barely big enough to hold them. The container must be big enough to permit the bearing to be sloshed around without touching the bottom where the dirt settles (Fig. 53). Never allow the bearings to touch bottom or they will be contaminated with dirt recently washed off.

Soak the bearings long enough to loosen the grease and dirt—several hours or overnight. Use a short bristled brush that will not lose its bristles or break off. After all visible dirt is removed, rinse the bearing in a container of *clean* solvent and then dip the bearing in oil.

If bearings may be reinstalled, inspect them carefully. When some doubt exists whether or not to replace the bearing, use this reasoning: if frequent inspections are given and the bearing is easy to replace, the risk of failure may not be so great; however, if infrequent inspections take place and the bearing is difficult to remove and install, the part should be replaced with a new one.

If the bearings have seals or shields, inspect them for damage and wear. If the seals are not removable and they are damaged or worn, the whole bearing must be replaced. Be sure to replace those seals that are replaceable if they are defective. Worn or damaged seals will allow dirt and moisture to enter the bearing and shorten bearing life.

Visually inspect the exterior of the bearing for cracks in the races, dented seals, and broken or damaged separators, balls, or rollers. If the bearing has been overheated, it will be a brownish blue or bluish black color. If any of these signs are found, the bearing should be replaced.

Inspect separable bearings for pitted, scratched, or flaked balls, rollers, or races. Replace the bearing if any of these signs of damage are found.

The inner surfaces and rolling elements in a suspected nonseparable bearing may be examined by shining a flashlight or reflected light from a strong light source between balls or rollers. Any visible pits or scratches are signs of damage and the bearing should be replaced.

Inspect outer race for proper wear pattern to assure no end loading or "tipping" has occurred.

If no visual signs of damage and wear are present, hold the bearing (Fig. 54) and rotate the outer race *slowly. Never spin it.* If any clicking or sticking is found, reclean the bearing. If after cleaning the condition still exists, replace the bearing.

Thrust bearings or tapered roller bearings can be inspected in the same way, except place the bearing on a clean surface and lightly apply hand pressure and rotate the bearing as shown at right in Fig. 54.

If the bearings are not to be reinstalled at once, wrap them in a clean, oil proof paper, place in a clean box and store in a dry, dust free place.

When bearings are found to be damaged, the cause must be determined or it may occur again when the new bearing is installed. This brings us to diagnosis of failures.

DIAGNOSING ANTIFRICTION BEARING FAILURES

Failures of antifriction bearings may be caused by several things, but keep in mind that properly maintained bearings do not wear out. They eventually fail due to fatigue. Failures for reasons other than *normal* material fatigue are considered to be premature failures.

Fatigue Failures

The first indications of *fatigue failure* may be noisy running and increased vibration. Upon examination of the bearing races you will find that the metal surfaces of the races have flaked away as shown in Fig. 55. This flaking is caused by the effects of speed and load.

ROTATE STRAIGHT BEARING
OUTER RACE SLOWLY

X2928

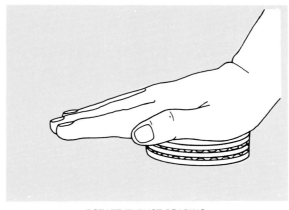

ROTATE THRUST BEARING
BY LIGHT HAND PRESSURE

Fig. 54 — Inspecting Bearings

Fig. 55 — Fatigue Failure Of Bearing

A decrease in speed or load will extend bearing life. Conversely, an increase in speed and load will shorten bearing life. This is why it is important to use the correct bearing for each specific application. Otherwise, premature fatigue failure may occur.

Premature Failures

Premature failures are usually the result of one or more of the following causes:

1. *Contamination*
2. *Distortion*
3. *Misalignment*
4. *Improper fitting*
5. *Improper lubrication*
6. *Vibration while bearing is stationary*
7. *Electrical current passing through bearing*
8. *Defects in material*
9. *Improper servicing techniques*

Note: Ball, roller, and needle bearings are subject to similar failures that have the same causes and effects.

Contamination

Contamination in a bearing is any foreign matter that will cause damage to the bearing. Moisture and any type of abrasive, such as dirt or sand will cause premature failure of the bearing.

SCRATCHES   RUST

Fig. 56 — Contamination Caused This Damage

Fig. 56 shows damage caused to bearing races by abrasive contaminants and moisture. The abrasives scratched and scored the race. Moisture caused the other bearing race to rust. Both types of damage, of course, cause premature failure. This type of damage may be prevented by using the correct lubricant, keeping bearings clean while handling them, and using new or undamaged seals.

Rust damage may also appear after extended storage or nonuse periods. This is mainly due to condensation. Prevent this by placing a light coat of petroleum jelly on the bearing when it is placed in storage or by storing bearings in a container of oil.

Fig. 57 — Distortion Caused This Premature Failure

Distortion

Distortion of the shaft or housing bore may cause premature fatigue failure (Fig. 57). If the shaft or housing is out of round, the rolling elements will be squeezed where there is the least clearance. This increased pressure will cause the race surfaces to flake as happens with normal fatigue. If the housing is out of round, regrinding it may solve the problem. The same is true with the shaft. If either cannot be salvaged, then replace the defective parts.

Misalignment

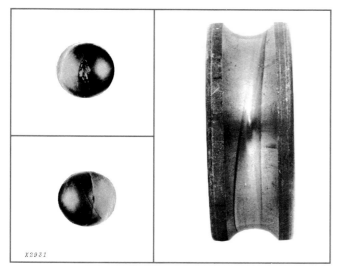

X2931

Fig. 58 — Misalignment Caused This Damage

Misalignment can be caused by a bent shaft, shaft shoulder not square, nonparallel housing bores, out of line housing bores, and foreign matter between the bearing and its seat. The typical effects of misalignment are shown in Fig. 58. Notice the paths worn in the races and the worn shape of the balls.

Misalignment in roller or needle bearings usually results in extreme pressure on the races and rollers which will result in premature fatigue failure. Determine the cause of failure and correct it or the same damage will occur when new bearings are installed.

Improper Fitting

Improper fitting will also cause premature bearing failure. Shown in Fig. 59 are three types of damage resulting from improper fitting.

The first example is a split race caused by forcing the bearing onto a shaft too large for the inside diameter of the bearing race, or by having the bearing "cocked" during installation.

The next example is the outer race of a bearing that was damaged by "fit rust" or fretting corrosion which results when the outer race fits too loosely in its housing. Do not sand or remove this rust. This will only increase the loose fit. Replace the bearing.

Creep wear, shown in the third example, is caused by a loose fit between the shaft and the inside bore of the bearing. As this wear progresses, the inner race will turn faster which will create more friction and heat leading to eventual bearing failure.

Improper Lubricant

In Fig. 60 you can see why the use of proper lubricant is so important.

The example shows smeared metal. This was caused by the rolling elements sliding across the surface rather than rolling. Too much lubricant that was too solid was used in this bearing. The thick consistency of

SPLIT RACE   FIT RUST   CREEP WEAR

X2932

Fig. 59 — Improper Fitting Caused These Failures

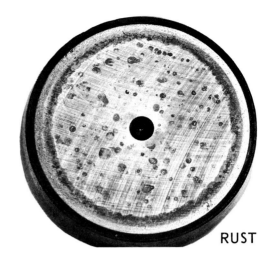

SMEARING   RUST

Fig. 60 — Incorrect Lubricant Caused This Damage

the lubricant retarded the rolling elements enough to start the sliding. As the metal surfaces became smeared, the rate of wear increased. This damage can also result from inadequate lubricant.

In the second example (at right) the bearing surface rusted. Some lubricants have moisture in them that will cause rust. The same problem will develop if moisture enters the bearing because of a defective seal. Rust will mix with the lubricant and form an abrasive compound which will cause rapid wear.

Always use the lubricant specified by the manufacturer.

Vibration Damage

Fig. 61 — Vibration Damage To Bearing

Most antifriction bearings are rolling while under load. The bearing pictured in Fig. 61, however, was stationary while subjected to *vibration*. The depressions are the result of a combination of wear and impacts caused by this vibration.

Fig. 62 — Electrical Current Damaged This Bearing

When this bearing is subjected to a rolling load, it will fail rapidly.

Electrical Current Damage

When bearings are in the presence of electricity, damage can occur if the *electrical current* passes through the bearing.

Fluting occurs (Fig. 62) as the bearing turns and electrical arcs melt the surface metal.

The cause of the electrical leak must be found and corrected or the electricity will also damage the new bearing.

Defects in Bearing Material

Because of excellent quality control, few *defects in material* or manufacturing are found in bearings. However, defects are exceptions that still occur occasionally.

Some of these manufactured "damages" are shown in Fig. 63 on the next page. Often, only a metallurgist can distinguish these types of defects versus the premature failures we just described.

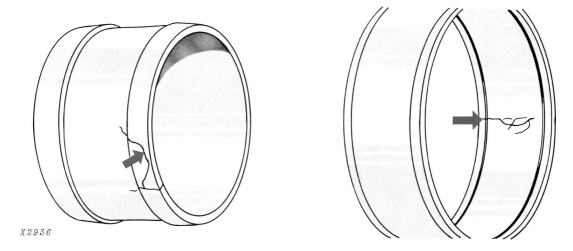

X2936

Fig. 63 — Defects In Bearing Material

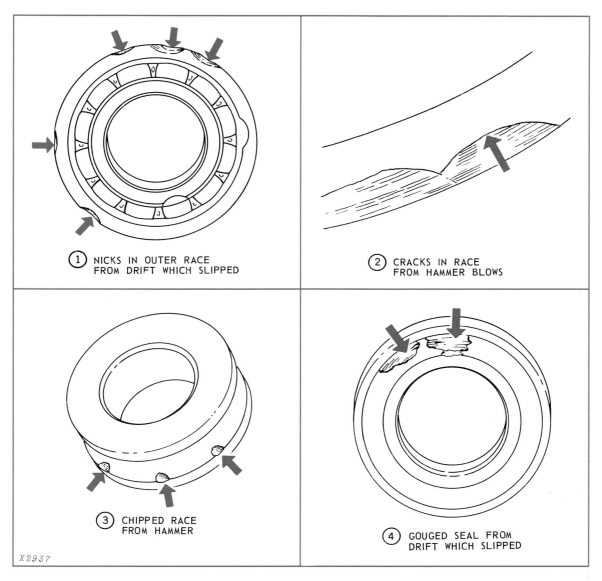

① NICKS IN OUTER RACE
FROM DRIFT WHICH SLIPPED

② CRACKS IN RACE
FROM HAMMER BLOWS

③ CHIPPED RACE
FROM HAMMER

④ GOUGED SEAL FROM
DRIFT WHICH SLIPPED

X2937

Fig. 64 — Damage Caused By Incorrect Service Techniques

Improper Service

If *improper service techniques* are used to remove or install bearings, damage may result which will cause bearing failure.

Fig. 64 on the preceding page illustrates some examples of damaged bearings that were serviced improperly.

Example No. 1 shows nicks in an outer race caused by using a drift to drive the bearing.
Example No. 2 is of cracks caused by striking the race with a hammer.

The results of this hammering also damages the inner race as shown in Example No. 3. When the race is struck a severe blow, this force is transmitted through the rolling elements to the other race and causes the chipping shown.

Another serious result of using improper tools is the damaged seal shown on the sealed bearing in No. 4. A drift slipped and gouged this seal. The seal's effectiveness is reduced and the separator is probably put in a bind. Both problems may lead to premature failure of the bearing.

See the "Removal" and "Installation" techniques described elsewhere in this chapter for proper servicing techniques.

LUBRICATION OF ANTIFRICTION BEARINGS

Antifriction bearings do not normally wear out. They fail because of fatigue. The flexing, loads, and heat eventually cause the metal to lose its strength. Of course, antifriction bearings will wear out if they are not lubricated properly.

A lubricant must perform three functions:

1. *Maintain a film of lubricant between all elements of the bearing to prevent rubbing.*

2. *Reduce friction between balls or rollers and the separators.*

3. *Prevent rust by coating all parts with a protective coating.*

Methods of Lubrication

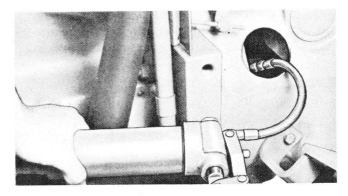

Fig. 65 — Applying Grease To Pressure Fitting Using Hand Gun

In addition to applying lubricant to bearings with a grease gun as shown in Fig. 65, packing and splash methods are also used. Keep in mind that the design and application of the bearing determines the type of lubricant required and the frequency of lubrication. Always follow the manufacturer's recommendations. He knows what the requirements are to insure long bearing life.

Fig. 66 — Packing A Bearing By Hand

Some bearings, such as wheel bearings, require infrequent lubrication. These bearings are usually packed with grease by either using a grease packer or by hand packing as shown in Fig. 66.

Remember that too much lubrication can be as harmful as too little. If too much grease is applied to a bearing, churning will cause friction and heat, leading to premature bearing failure. Again, follow the manufacturer's instructions.

X2939

Fig. 67 — Properly Packed Ball Bearing

As a general rule, *pack antifriction bearings half full.* Remove any excess lubricant. In Fig. 67 you can see a properly packed ball bearing. With both ball and roller bearings, pack both sides even with the cages. Then rotate the races a few times to remove excess grease.

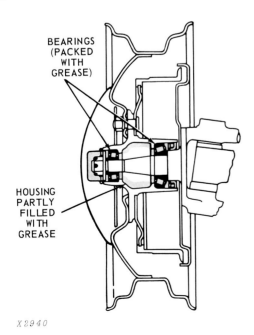

BEARINGS
(PACKED
WITH
GREASE)

HOUSING
PARTLY
FILLED
WITH
GREASE

X2940

Fig. 68 — Partially Fill Housing With Grease

Before installing the bearings in the housing, partially fill the housing with grease (Fig. 68). Do not completely fill the housing. This will cause churning of the grease and will result in heat and eventual bearing failure. Add only enough grease to prevent thinned grease from running out of the bearings.

In splash systems of lubrication, such as in transmissions, the most important concerns are:

1. *Keep sufficient lubricant in the housing.*

2. *Use the correct lubricant.*

3. *Change lubricant at recommended intervals.*

4. *Change filter at recommended intervals.*

5. *Use lubricant rated for proper temperature. Higher temperatures than recommended for a given lubricant will break down extreme pressure additives.*

Usually, filler plugs or caps provide a means of filling the housing.

Note: For more on lubricants, see the FOS manual on "Fuels, Lubricants, and Coolants."

INSTALLATION OF ANTIFRICTION BEARINGS

Proper installation is important if bearing damage is to be prevented. Fig. 69 on the next page shows several *acceptable* methods that can be used to install bearings without damaging them. (Fig. 70 shows the *wrong* methods.)

The first examples (1) show two proper methods of pressing an inner race on a shaft. Notice that no pressure is exerted on the outer race. When pressing the outer race into a housing (2), apply pressure to the outer race only as shown in the next example.

A push-puller (3) may be used in certain situations to install a bearing. When care is taken, a race may be driven onto a shaft with a hammer and tube as shown in (4).

Needle bearings (5) require a special driver which is undercut as shown by the inset drawing. This undercut is shaped to transmit the driving force to the *outer shell*, preventing the bearing shell from buckling.

 CAUTION: Oil vapor may catch fire when heated or near a flame; be careful.

Some bearings have a very tight fit on the shaft and must be heated for ease in assembly. If the bearing is small, it can be heated in a container of oil; but, the temperature should not exceed 300° F (149° C). The bearing must not contact the bottom of the container because it will absorb the heat directly and may be damaged. Remember that bearings are heat treated to harden them. If the bearing gets too hot, it will lose some of its hardness.

A better method of heating bearings for installation is by inserting a light bulb into the bearing bore (if the bore is large enough). Use a cover over the bearing

and bulb to retain the heat until the bearing is ready for installation. This method is less hazardous to both the bearing and you. The fire danger is eliminated. The bearing damage due to high temperatures is eliminated also.

Bearings may be installed haphazardly and result in bearing damage. Many of these *wrong* methods are shown in Fig. 70.

Always press on the bearing race that has the press fit—never on the slide-fit race as shown in (A) and (D), Fig. 70.

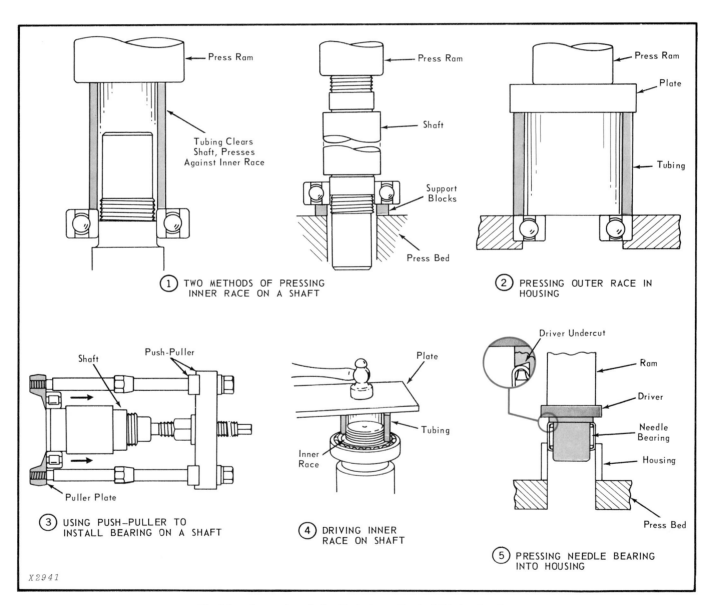

Fig. 69 — Correct Installation Techniques For Anti-Friction Bearings

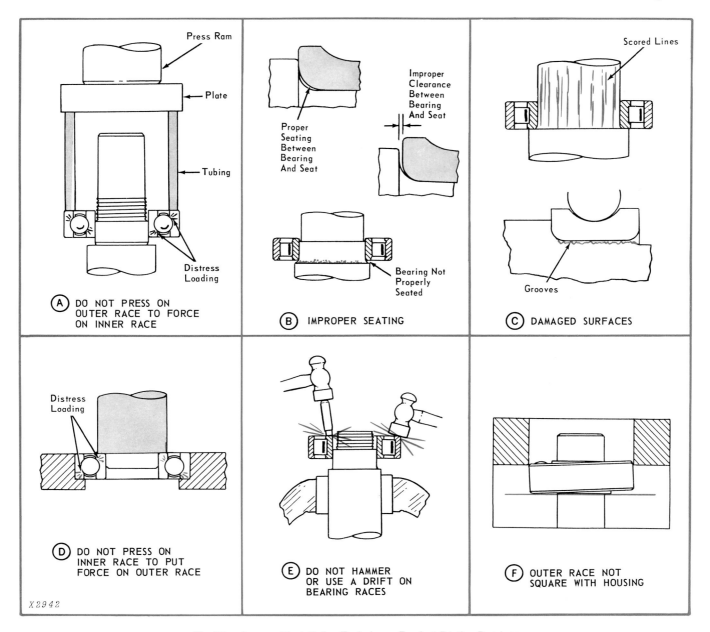

Fig. 70 — Incorrect Installation Techniques For Anti-Friction Bearings

Be sure the bearing is properly seated on the shaft or in the housing (B). Some bearings have sharp edges and will not seat if a fillet is present.

Also, be certain that the shaft or housing is clean and free of dirt or metal particles (C).

This foreign matter can prevent the bearing from seating completely. A bearing that is not seated properly may be subjected to high preload which will cause premature failure.

A damaged shaft or housing surface may also lead to short bearing life. Scored or grooved surfaces (C) will peen down under load and the bearing race will turn

because clearance is increased. As described earlier, when bearing creep occurs at a rapid rate, friction and heat increase, causing bearing failure. Do not hammer or use a drift on bearing races (E). This will damage the races and the rolling surfaces; see "Diagnosis of Failures."

Be certain that the outer race is square with the housing when installing bearings (F). A bearing can be severely damaged if it is cocked while being pressed into a housing.

Now that we have looked at installation techniques, let's talk about adjustments.

ADJUSTMENT OF BEARINGS

Many antifriction bearings are mounted so that no adjustment is required. Usually bearings that do not require adjustment are mounted with one bearing "floated" and the other secured by one of several means.

Those bearings that do require adjustment are usually fixed in place by securing them with one of the devices described in "Mountings." This adjustment is called *preloading*.

Preloading is used to reduce deflection and provide proper running conditions where dynamic loads are present.

In instances where preloading is specified, care must be taken not to set the preload too tight. This will cause heat to build up and the bearing will fail. If too little preload is used, the supported parts will deflect excessively and rapid wear will result.

Because preload cannot be measured, it has to be established by first creating endplay in the assembly. This is usually done by adding or removing shims, depending on design. Once endplay is established it has to be measured. If the endplay, for instance, is 0.005 inch (0.13 mm) and the preload is to be 0.002 inch (0.05 mm), then a shim thickness of 0.007 inch (0.18 mm) has to be removed or added, depending on the design. The best results are achieved if the assembly is in the vertical position during this procedure. Some manufacturers also recommend that correct preload be measured by the "rolling torque" method. In this procedure the assembly is rotated and the resistance is measured with a torque wrench or a spring gauge.

Consult Technical Manual or Service Manual specifications for the proper preload for each bearing you install.

To prevent serious damage to components using antifriction bearings, it is important to know what symptoms indicate trouble. This brings us to trouble shooting.

TROUBLE SHOOTING OF ANTIFRICTION BEARINGS

Properly maintained antifriction bearings should be relatively trouble free. Use the following chart to determine cause and remedy of any bearing failure symptoms.

If the suggested remedy does not correct the problem, then disassemble the bearing and determine the cause.

Refer to "Diagnosis of Failures" to help identify failures.

BEARING REPLACEMENT GUIDELINES

All bearings are stamped or coded with the manufacturer's identification. A bearing compatibility chart is required if a bearing from one manufacturer is replaced by another. Bearings that look like the original may not be a suitable replacement. Bearings must be compatible to the machine:

• Oils and greases
• Fit
• Surface
• Load carrying ability

Bearings that do not meet these standards of compatibility may be a hazard to the machine and the operator.

Machines generally are designed with the following guidelines for bearing selection.

1. Ball bearings are used for lighter loads.

2. Roller bearings are used for heavier loads.

3. Roller bearings work better under shock or impact loading.

4. Shaft misalignment between the bearing usually requires self-aligning ball or spherical roller bearings.

5. Ball thrust bearings are designed for thrust loads in one direction only.

6. Angular contact ball bearings or deep groove ball bearings are better for high speed thrust loads.

7. Low friction applications usually call for self-aligning ball bearings and cylindrical roller bearings.

8. Deep groove ball bearings can be sealed when bearings and machine surfaces can be contaminated.

Bearing types must never be mixed unless there is specific information concerning their compatibility.

Possible Causes	Symptoms				Remedy
	Excessive Noise and Vibration	Excessive Heat	Lubricant Loss	Shaft Resistant To Turning	
Insufficient lubrication	X	X			Fill with correct amount of proper lubricant.
Incorrect lubricant	X	X		X	Use recommended lubricant.
Defective bearing	X				Replace bearing.
Dirt	X			X	Clean bearing. Replace seals and bearing if damaged.
Corrosion	X			X	Use lubricant that resists corrosion. Replace bearing.
Improper preload	X	X		X	Adjust to specification.
Misalignment	X	X			Determine what is misaligned: housing, shaft, or bores. Shaft may be bent or housing bores not aligned. Replace shaft or housing.
Bearings fit too loose	X				Chrome plate or metalize shaft or bore and regrind to specification or replace bearing if seating surfaces are worn.
Improper mounting	X				Bearing mounted crooked or put in a bind causing excessive preload.
Improper lubricant			X		Use high-temperature grease.
Lubricant churning		X			Use less lubricant or use a lower-viscosity lubricant. Be sure to use recommended lubricant.
Excessive creep		X			Either inner or outer race is sliding around its seated surface. Build up shaft or housing or replace bearing, whichever is applicable.
Leakage through seal			X		Replace seal. Check shaft for scratches or burrs that may have damaged seal.
Lack of lubrication				X	Add proper lubricant. Check for damage.
Seal tight				X	Use proper size and type of seal.
Bearing cocked				X	Remove bearing and clean all mating surfaces. Reinstall bearing if not damaged.
Leakage between bearing caps and housing			X		Replace gasket.

NOTE: Additional information on bearing failures can be found in F.O.S. Identification of Parts Failures, Chapter 8.

TEST YOURSELF

QUESTIONS

1. Name three of the four main functions of bearings.

2. What are the two load forces concerning bearings?

3. Regarding plain bearings, what state of lubrication is the most desirable?

4. Why is bearing crush important?

5. What determines the load carrying capacity of antifriction bearings?

6. A full ball bearing was used as a replacement for a Conrad bearing; the full bearing soon failed. Why did it fail?

7. You are installing antifriction bearings in a bearing housing and on a shaft that turns. What race of the bearing is going to be the press fit?

8. You are going to press a ball bearing onto a shaft. What must you be careful *not* to do?

9. If an antifriction bearing fails due to normal use, what is the failure called?

10. If an antifriction bearing suffers fatigue failure prematurely, what is the most likely cause?

11. How is preload determined?

(Answers in back of text.)

SEALS / PART 2

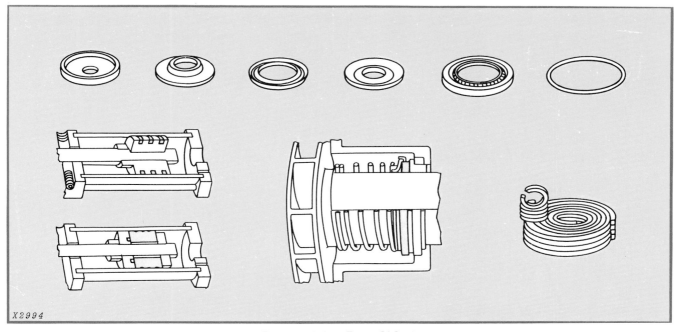

Fig. 1 — Various Types Of Seals

TYPES OF SEALS

All modern equipment uses **seals** to keep in fluids, or gases, keep out dirt and moisture, and sometimes to hold pressures or vacuums.

There are two basic types of seals:

• **Dynamic — to seal moving parts**

• **Static — to seal fixed parts**

DYNAMIC SEALS (Fig. 2) include shaft and rod seals, compression packings, and piston rings.

STATIC SEALS are usually gaskets, but may also be O-rings or packings.

Another category of static seals are **sealants,** which are similar to gaskets except that they are *applied* as a liquid or paste. These sealants will cure to a solid or semi-solid state.

Listed in the box on the next page are all the seals we will cover in this chapter.

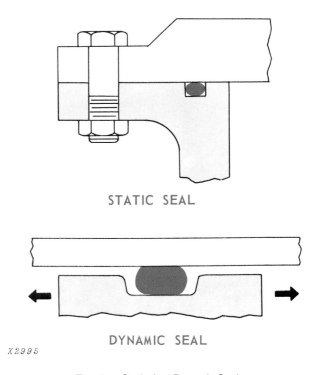

STATIC SEAL

DYNAMIC SEAL

Fig. 2 — Static And Dynamic Seals

DYNAMIC SEALS	STATIC SEALS
Radial Lip Seals	Nonmetallic Gaskets
Exclusion Seals	Metallic Gaskets
Clearance Seals	Static O-Rings
Ring Seals	
Face Seals	**SEALANTS**
Compression Packings	Hardening Types
Molded Packings	Nonhardening Types
Diaphragm Seals	Tapes

The perfect seal should prevent all leakage. However, this is not always practical or even desirable. In dynamic uses, for example, slight leakage as a fluid *film* aids in lubrication and cooling of moving parts. The point to keep in mind is that each type of seal has a different function.

Seals appear to be simple objects when held in the hand. But in use they are complex, precision parts and must be treated carefully if they are to do their jobs properly. Remember, seals are a vital defense against leaks at the critical openings or joints in all machines.

HOW SEALS ARE SELECTED

The designer of a system has many factors to consider when choosing a seal. Some of these are:

1. *Will the seal resist all pressures expected?*

2. *Can the seal withstand the heat of operation?*

3. *Will the seal wear out too rapidly?*

4. *Will the seal be harmed by the fluid or gas?*

5. *Does the seal fit without dragging on the moving parts?*

6. *Will the seal score or scratch polished metal parts?*

Each application for a seal presents a different set of problems. This is why so many seals are on the market today. And why the right choice of a seal for replacement is so important. Always follow the manufacturer's recommendation.

Remember that all seals are identified either by a part number on the case or by other markings (as paint slashes on O-rings). These identifications tell you if the seal is meant for your application, if it can resist the temperatures, the chemicals in the system fluid, etc.

DYNAMIC SEALS

In this section we will cover all the **dynamic** seals listed in the box at left.

RADIAL LIP SEALS

Radial lip seals are used primarily for keeping lubricants in systems which have moving shafts.

These seals are also called *oil seals* or *shaft seals*.

A typical lip seal is shown in Fig. 3. All major elements are named (SAE nomenclature). The basic parts are the outer metal case and the inner flexible sealing element.

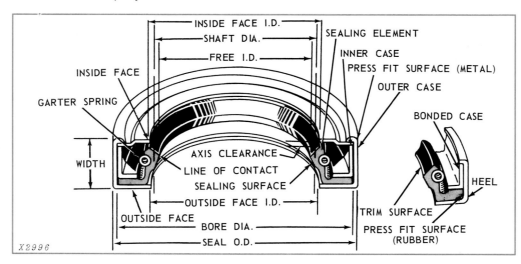

Fig. 3 — Radial Lip Seals

ADVANTAGES

Basic advantages of lip seals are:

1. *Fit into small space*

2. *Fairly low cost*

3. *Easy to install*

4. *Handle many variables while sealing*

Variables include most oils and hydraulic fluids over wide temperature ranges, moderate fluid pressures, average misalignment, dynamic shaft runout, and variations in shaft speed.

BONDED SEAL

ASSEMBLED SEAL

X2997

Fig. 4 — Basic Types Of Radial Lip Seals

TYPES

Two basic types of radial lip seals are:

- **Bonded seals**

- **Assembled seals**

In the BONDED seal (Fig. 4), the flexible sealing element is formed and bonded to the metal case in one operation.

In the ASSEMBLED seal (Fig. 4), the parts are made separately and assembled into the outer case, which is then crimped or rolled.

OPERATION

A lip seal seals because of an interference fit between the flexible sealing element and the shaft. Usually spring pressure behind the lip is also added. Keeping the fluid in depends upon a precise amount of lip contact pressure.

The seal lip should ride on a thin film of lubricant.

The film does the sealing so it must be controlled precisely by the mechanical pressure of the sealing element and the shaft finish.

If the fluid film gets too thick, fluid leaks; if too thin, the lip wears and friction, heat, and lip oscillations can result.

Normally, if contact pressure increases, the film gets thinner. Heat build up can also reduce the fluid film.

Never run lip seals without lubrication unless necessary, and then only for short periods.

Unless defective or installed wrong, lip seals should only leak the thin fluid film we discussed above.

SELECTION OF LIP SEALS

Radial lip seals are classified by lip types as follows:

- ***Single Lip*** *(Fig. 5, A). Lip not spring loaded. For containing viscous fluids like grease at slow shaft speeds.*

(A) SINGLE LIP SEAL
Not Spring-Loaded

(B) SINGLE LIP SEAL
Lip Spring-Loaded

(C) DOUBLE LIP SEAL
One Lip Spring-Loaded

(D) DUAL LIP SEAL
Both Lips Spring-Loaded

X2998

Fig. 5 — Basic Lip Types Of Radial Lip Seals

- **Single Lip Spring Loaded** (Fig. 5, B). For retaining lower viscosity lubricants at higher shaft speeds in clean atmospheres.

- **Double Lip** (Fig. 5, C). Lips face opposite, one spring loaded, one not. For retaining lubricant on spring loaded side, while keeping out dirt on the other side.

- **Dual Lip** (Fig. 5, D). Lips face opposite, both are spring loaded. For containing lubricant on one side and excluding fluid on the other.

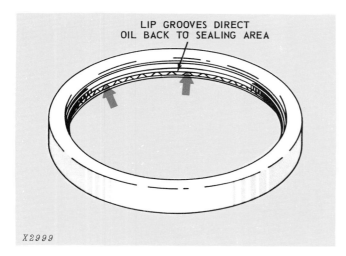

LIP GROOVES DIRECT
OIL BACK TO SEALING AREA

X2999

Fig. 6 — Hydrodynamic Seal

Other special lip seals include:

1. **Split seals** for difficult installations.

2. **External seals** for a fixed shaft and rotating bore (outer part of seal is the sealing surface).

3. **Hydrodynamic seals** (Fig. 6) which have grooves or ribs molded into the seal lip to direct oil flow back into the sealing area, reducing heat and wear. This seal can operate at less lip pressure than standard lip seals. It works best for one direction rotation at fairly high surface speeds.

LIP SEAL FAILURES

Lip seals have sealing elements of rubber or synthetics which require special care. Metal cases can also be bent or "cocked." Some of the major service problems and procedures are given below.

Checking Seals For Leakage

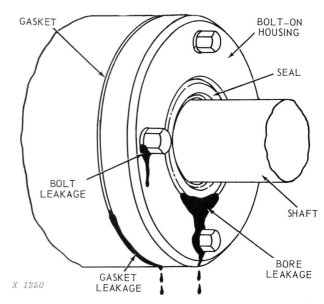

GASKET

BOLT-ON HOUSING

SEAL

BOLT LEAKAGE

SHAFT

GASKET LEAKAGE

BORE LEAKAGE

X 1250

Fig. 7 — Common Types Of Oil Leaks

Before disassembly, check out the causes of leakage. This may prevent a repeat seal failure, caused by problems other than the oil seal.

Before cleaning the area around the seal, find any paths of leakage (Fig. 7). Sometimes leakage may be from sources other than the seal: worn gaskets, loose bolts, cracked housings, or loose line connections.

Inspect the sealing area of the seal to see if it is wet or dry. If wet, see whether the oil is running out or is merely a lubricating film.

Removing Seals

During removal, continue to check for causes of leakage.

Check both the inner and outer parts of the seal for wet oil which means leakage.

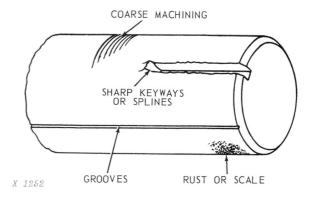

X 1252

Fig. 9 — Shaft Conditions Which Can Damage Seals
And Cause Leakage

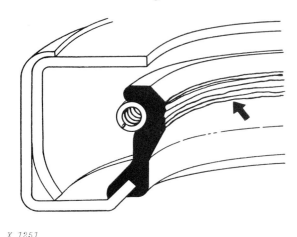

X 1251

Fig. 8 — Seal Worn By Rough Shaft

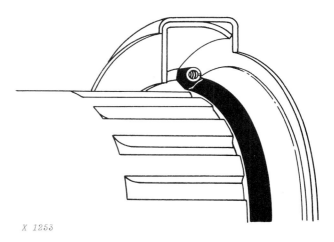

X 1253

Fig. 10 — Shaft Splines Or Keyways Can Damage Seals
During Installation

Find out if a shaft spline, keyway, or burred end could have caused a nick or cut in the seal lip during installation (Fig. 10).

When removing the seal, inspect the sealing lips (Fig. 8) before washing them. Look for unusual wear, warping, cuts and gouges, or particles embedded in the seal.

On spring loaded lip seals, be sure the spring is seated around the lip, and that the lip was not damaged when first installed.

Normally, do not disassemble the unit any more than necessary to replace the faulty seals.

Checking Shafts and Bores

Check shafts for roughness at seal contact areas (Fig. 9). Look for deep scratches or nicks that could have damaged the seal.

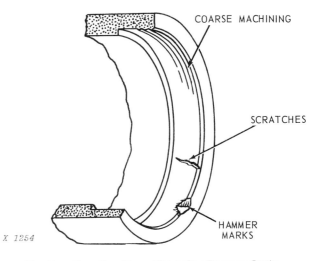

X 1254

Fig. 11 — Bore Conditions Which Can Damage Seals
And Cause Leakage

Inspect the bore where the seal is pressed (Fig. 11). Look for nicks and gouges that could create a path of oil leakage. A coarsely machined bore can allow oil to seep out by a spiral path. Sharp corners at the bore edges can score the metal case of the seal when it is pressed in. These scores can make a path for oil leakage.

Some engines contain replaceable wear sleeves on the crankshaft (Fig. 12). The sleeve can be easily removed by splitting it with a sharp chisel. Don't nick the shaft.

Before installing a new wear sleeve, check the shaft for nicks or scratches from improper handling.

Apply a sealing compound on the inside of a new sleeve. Then install with the proper tools to avoid damage to the sleeve.

Fig. 12 — Replaceable Wear Sleeve

Checking Seals For Compatibility With Fluids Or Operating Temperatures

Some lubricants are harmful to certain seals, especially rubber lips. An incorrect oil can either harden or soften the synthetic rubber in seals and so damage them.

If the seal lip is "spongy," this probably means that the seal and the fluid are not compatible. If the seal is factory approved, then an improper fluid may have been used in the system.

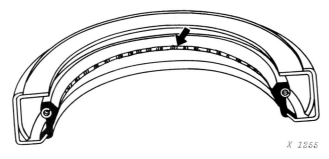

Fig. 13 — Seal Lips Damaged By Heat

A hard seal lip can be caused by either heat or chemical reaction with an improper fluid.

A hard seal lip at the point of shaft contact (Fig. 13) is generally the result of *heat* from the shaft or fluid, or from running the seal dry.

INSTALLING LIP SEALS

To get the best use of seals, proper handling and installation are vital. Lip seals are fragile and can be easily damaged.

To prevent this, keep seals protected in their containers until ready for use. Store them in a cool, dry, clean place. Seals should be given the same care as precision bearings.

As a general rule, **replace all seals that are disturbed during repair of the system.** The price of a few seals is economical compared to a return repair job to correct leaks.

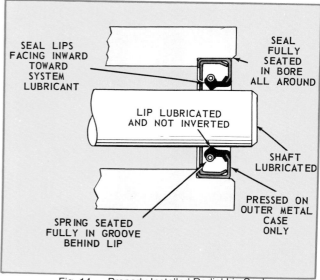

Fig. 14 — Properly Installed Radial Lip Seal

Here are some rules for seal installation:

1. Install seals recommended by the manufacturer of the machine.

2. Use the fluids recommended in the machine operator's manual.

3. Keep the seals and fluids clean.

4. Before installing seals, clean the shaft or bore area. Inspect these areas for damage. File or stone away any burrs or bad nicks and polish with a fine emery cloth for a smooth finish, then clean the area to remove metal particles. In dynamic applications, the sliding surface for the seal should usually have a mirror finish.

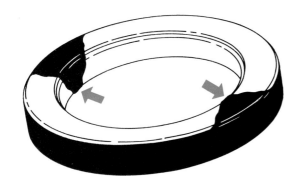

Fig. 15 — Too Much Sealant On Case Runs Onto Sealing Lips

5. On seals with single lips, the lip should normally face in **toward** the system lubricant. This is usually the garter spring side.

6. Lubricate the seal, especially the lips, and the shaft to ease installation. Use the system fluid to lubricate the seal and shaft.

7. With metal cased seals, coat the seal's outside diameter with a **thin** film of gasket cement to prevent bore leakage. Do not allow excess cement to run onto sealing lips (Fig. 15).

NOTE: Precoated seals do not require cement on the bore fit or the outer case of the seal.

8. Use the recommended tool to install the seal properly (Fig. 16, No. 1 and 4). This is very important with pressed-in seals. If a seal driving tool is not available, (1) use a circular ring such as an old bearing race that contacts the seal case near the outer diameter, or (2) use a square wooden block. **Do not use sharp tools, and never press on the sealing lip; only the outer metal case.**

9. Use a cone or sleeve to protect the sealing lip when installing the seal over sharp corners of shafts, keyways, or splines (Fig. 16, No. 2, 3).

10. Shim stock can also be used to protect seals when installing them over sharp edges such as shaft splines. Place *rolled* plastic shim stock 0.003-0.010 inch (0.076-0.25 mm) over the sharp edge, then pull it out after the seal is in place.

11. Be sure the seal is driven in squarely to prevent "cocking" the seal (Fig. 17 on next page). A cocked seal allows oil to leak out and dirt to enter. **Be careful not to bend or "dish" the float metal area of metal cased seals. This also distorts the lips.**

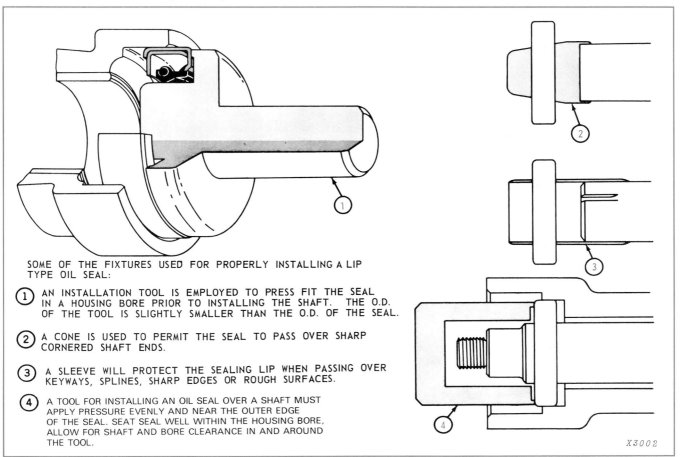

SOME OF THE FIXTURES USED FOR PROPERLY INSTALLING A LIP TYPE OIL SEAL:

① AN INSTALLATION TOOL IS EMPLOYED TO PRESS FIT THE SEAL IN A HOUSING BORE PRIOR TO INSTALLING THE SHAFT. THE O.D. OF THE TOOL IS SLIGHTLY SMALLER THAN THE O.D. OF THE SEAL.

② A CONE IS USED TO PERMIT THE SEAL TO PASS OVER SHARP CORNERED SHAFT ENDS.

③ A SLEEVE WILL PROTECT THE SEALING LIP WHEN PASSING OVER KEYWAYS, SPLINES, SHARP EDGES OR ROUGH SURFACES.

④ A TOOL FOR INSTALLING AN OIL SEAL OVER A SHAFT MUST APPLY PRESSURE EVENLY AND NEAR THE OUTER EDGE OF THE SEAL. SEAT SEAL WELL WITHIN THE HOUSING BORE, ALLOW FOR SHAFT AND BORE CLEARANCE IN AND AROUND THE TOOL.

X3002

Fig. 16 — Tools For Properly Installing Lip-Type Oil Seals

12. After assembly, always check the unit by hand for free operation.

13. Try to prevent dirt and grit from falling on shafts and being carried into the seal. This material can quickly damage the seal or score the metal surfaces.

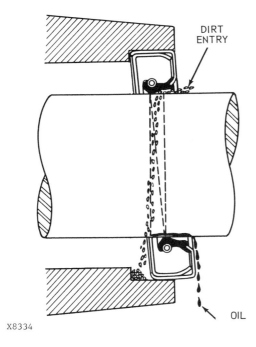

X8334

Fig. 17 — Cocked Seals Allow Dirt To Enter And Oil To Leak Out

Run-In Checking Of New Lip Seals

When a new lip seal is installed on a clean shaft, a break-in period of a few hours is required to seat the seal lip with the shaft surface. During this period, the seal polishes a pattern on the shaft and the shaft in turn seats the lip contact, wearing away the knife sharp lip contact to a narrow band.

During this period, slight seepage may occur. After seating, the seal should perform without any measurable leakage.

EXCLUSION SEALS

Exclusion seals are used to keep foreign material out of the moving parts of machinery. This protection is necessary because dirt can contaminate the lubricant and accelerate wear and corrosion.

Static joints are easily sealed by tight fits and gaskets. Sealing between parts having relative motion, such as between a housing and a moving shaft, is more difficult.

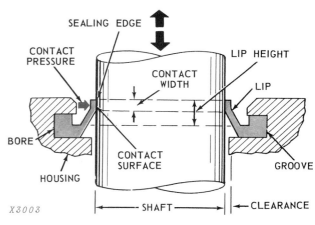

X3003

Fig. 18 — Radial Exclusion Seal

Never use seals designed for inclusion for excluding applications, except for very light service. They usually do not perform properly and also are damaged by even small amounts of abrasive material.

Exclusion seals can be classified into five general groups:

- **Radial seals**
- **Wipers**
- **Scrapers**
- **Axial Seals**
- **Boots**

Fig. 18 is a diagram of a typical radial exclusion seal.

WIPERS

Exclusion seals with lips of leather, elastic, or similar material are commonly called wipers when they are used on reciprocating shafts (see Fig. 18). Many of these can also be used on rotating shafts. These seals resemble radial lip seals.

The active sealing element or lip is usually a part extending outward and contacting the shaft. Sufficient radial or contact pressure must be provided to keep the contact surface against the shaft. Pressure is supplied by the band or cantilever action of the lip or by auxiliary springs.

Lip Design

A good inclusion seal will allow a slight film of fluid to pass the sealing edge to lubricate the contact surface. The wiper, however, must stop all material at the sealing edge because the medium it seals is seldom lubricating and is often abrasive.

The major consideration in wiper lips is that the sealing edge be kept in intimate contact with the shaft.

The lip is normally short and sturdy, particularly for reciprocating shafts, to overcome the resistance of the foreign material. At the same time, the lip must be flexible enough to follow shaft side motion.

Wiper Materials

Seal lips are usually leather or an elastic such as polyurethane or rubber.

SCRAPERS

Exclusion seals that have a metallic lip or scraping element are called *scrapers* (Fig. 19). They are used to scrape heavy or clinging material from reciprocating shafts. Wipers are sometimes used behind them to catch any fine particles or fluids which pass the scraper.

As the shaft moves inward toward the scraper, the sharp sealing or scraping edge of the scraper element removes foreign material. The scraping edge is kept sharp by the honing action of the shaft passing through it. Sufficient contact pressure to keep the edge in intimate contact with the shaft is supplied by prestressed spring action of the scraper element or by auxiliary springs.

Basically, there are two types of scrapers:

• **Conical Scrapers**

• **Scraper Rings**

These two types are shown in Fig. 19.

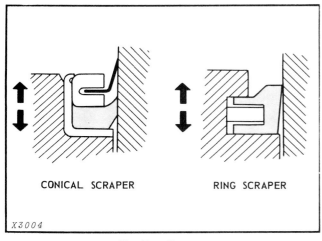

Fig. 19 — Scrapers

The CONICAL SCRAPER has a sharp knife edge which digs under the material and lifts it from the shaft.

The SCRAPER RING has a flat top as shown. Although this produces somewhat more resistance to scraping than the conical scraper, the scraping edge is quite sturdy.

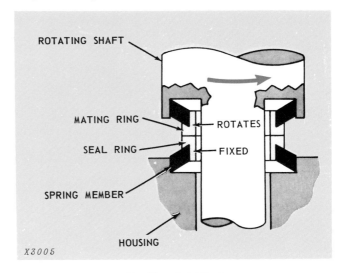

Fig. 20 — Axial Seal

AXIAL SEALS

Axial seals (Fig. 20) exert axial contact pressure to seal two flat surfaces having rotating motion with respect to each other.

Almost all axial seals are capable of exclusion and inclusion. Their use is usually confined to applications where the radial seal is inadequate.

All axial exclusion seals contain the same basic elements (Fig. 20). The seal shown could be considered as two identical mating seals.

The ring and mating parts are either both hard, or one is hard and the other soft. Contact faces of the hard rings are usually lapped to optical flatness, while those of the soft rings must only be flat enough to make good sealing contact under the available contact force. Hard rings are usually wrought, cast, sintered metals, or ceramics. Soft ring faces are leather, plastic, or elastics.

The most common axial seal is that of a water pump.

BELLOWS AND BOOTS

Since boots (or bellows) are used to prevent material from entering a mechanism, they are exclusion seals.

They differ from other seals in that they have no frictional contact. The motion is absorbed by the flexing of the boot itself.

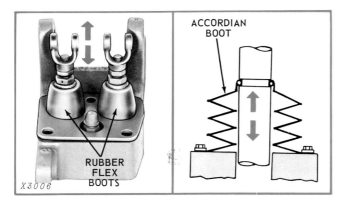

Fig. 21 — Boots Which Protect Reciprocating Shafts

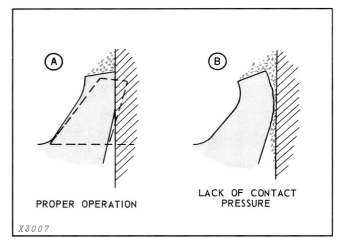

Fig. 22 — Lips Of Wiper Seals

Various shapes are made to cover and accommodate the motion of devices such as universal joints, ball joints, shift levers, and hinges (Fig. 21, left). These are made of sewn or cemented leather, molded elastics, or plastics.

The most common type is the accordion boot, which is used principally to seal reciprocating rods (Fig. 21, right).

MAINTENANCE OF EXCLUSION SEALS

Exclusion seals are built tough, but they will wear out early unless properly maintained.

Lubrication

Lubrication of the seal lip often is sparse. On reciprocating shafts, a slight film left by the inclusion seal is usually sufficient to prevent excessive wear and abrasion of the exclusion lip. But too thick a film will be scraped off the shaft by the exclusion lip. In rotation, the exclusion seal must depend upon seepage from the inclusion seal. Lubrication should be placed between the seals at installation. In some cases, a means of periodic lubrication is provided.

Temperature

Temperature is seldom high at the exclusion seal because it is located at the surface of the unit, where heat is rapidly dissipated, but temperature may become high with increased friction due to dirt.

Shaft Condition

Shaft condition is very important to life and operation of the seal. A medium shaft finish is recommended. Too fine a finish, however, will not hold sufficient surface oil film for lip lubrication, particularly if the surface is nonporous.

Lip Condition

Lip Condition is vital to the sealing of wipers (Fig. 22). If the lips start separating from the shaft as shown in (B), foreign material can enter and start a seal failure. The particles will scratch the shaft and wear out the seal lip. For this reason, the sealing lip must be kept in intimate contact with the shaft as shown in (A). If the lip starts to wear or weaken, replace the complete wiper.

CLEARANCE SEALS

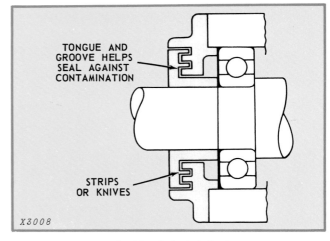

Fig. 23 — Labyrinth Seal

Clearance seals are dynamic seals which limit leakage by closely controlling the annular clearance between a rotating shaft and its relatively stationary housing.

The two basic types are:

• **Labyrinth**

• **Bushing (or ring)**

These seals are used when some leakage can be permitted, and pressure differentials are beyond the limits of contact seals (face and circumferential).

LABYRINTH SEALS

Labyrinth seals are used mainly in heavy industrial and power applications where relatively high leakage can be tolerated and simple design and reliability are a necessity.

A labyrinth consists of one of more thin strips or knives which are attached to either the stationary housing or the rotating shaft (Fig. 23).

Final running clearances are frequently established by contact between the knife and the housing, which rubs away or flares the knives.

Leakage of a compressible fluid is limited when the labyrinth causes the fluid to lose velocity pressure as it is throttled through the annular orifices.

Labyrinth sealing is especially effective when bearings operate in particularly dusty surroundings, as in grinding machines, and where bearings are splashed or subjected to large quantities of water as in paper-making machinery, turbines and centrifugal pumps.

Grease contributes to the efficiency of labyrinth sealing. It forms a protective layer in the housing adjacent to the labyrinth to catch any dust particles that may have penetrated the sealing device. The grooves of a labyrinth seal at assembly should be filled with a suitable sealing grease. When relubricating, grease is pressed into the labyrinth, forcing foreign particles out.

BUSHING AND RING SEALS

The bushing-type seal is a close-fitting stationary sleeve within which the shaft rotates.

Leakage from a high-pressure source at one end of the bushing to a region of low pressure at the other end is controlled by the restricted clearance between shaft and bushing.

Bushing and ring seals have a lower leakage per unit length than labyrinths.

Bushings are *fixed* or *floating* according to whether or not they are fixed with respect to the stationary housing.

RING SEALS

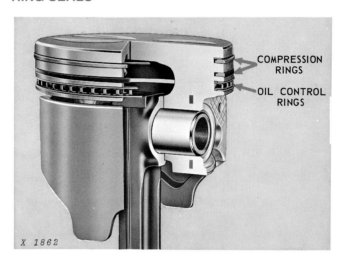

Fig. 24 — Split-Ring On An Engine Piston

Ring seals are divided into two classes:

• **Split-ring seals—for reciprocating parts**
• **Circumferential seals—for rotary parts**

Both types depend on surface contact between the seal and the moving part, and between the seal and the stationary part to perform their function.

There are several other similarities between the two types. For instance, in spite of the difference in names, both are composed of split rings.

However, the *split ring seals* are basically rings with a joint in each for ease of assembly. The rings on the engine piston in Fig. 24 are a good example.

The *circumferential seals* are more complicated in construction. Also, only some of the split rings require springs, while all the circumferentials use them.

Let's look at each type; first, the split ring.

SPLIT-RING SEALS

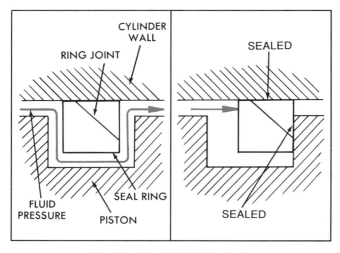

Fig. 25 — How A Split Ring Seals

Expanding split rings (piston rings) are used in compressors, pumps, and internal combustion engines.

Straight cut and seal joint rings are commonly used in industrial and aerospace hydraulic and pneumatic cylinders where rugged rings are needed and leakage can be tolerated.

Contracting split rings or rod seals are used in actuators where high pressure, high temperature, radiation, thermal fatigue, and reliability prevent use of packings.

Types

Ring seals are primarily sealed axially by the pressure of the medium to be sealed.

There are two main surfaces where a seal must be made: the I.D. of a cylindrical bore and the side of the piston for the expanding seals; and the O.D. of a rod and the side surface of a fixed housing or carrier for the contracting seals.

The way a split ring is brought into sealing position by pressure is shown in Fig. 25. In some seals, there are extra mating surfaces or joints in the ring itself that must be sealed.

STRAIGHT CUT SEAL RING

The simplest and least expensive type of split ring has a straight cut joint (Fig. 26, top).

Leakage through the joint is frequently of the linear type through a free passage. Hence, it depends on the gap dimensions, resistance of the passage, and the pressure and viscosity of the fluid.

The straight cut seal ring is used as a piston seal for lower grade actuators. Rarely is the type of ring suitable for a rod seal.

STEP SEAL RING

When free leakage is not permissible, a variety of positive sealing joints are used. The simplest of these joints is the step seal (Fig. 26, center). This seal has an overlapping surface on each half of the ring along an oblique angle. It seals simultaneously on the O.D. surface, the side surface, and the joint surface.

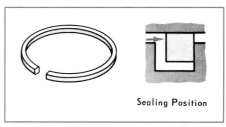

Sealing Position

STRAIGHT CUT SEAL RING

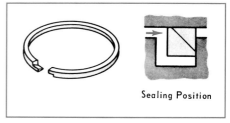

Sealing Position

STEP SEAL RING

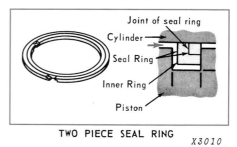

Joint of seal ring
Cylinder
Seal Ring
Inner Ring
Piston

TWO PIECE SEAL RING   X3010

Fig. 26 — Types Of Ring Seals

TWO PIECE SEALING RING

To seal in both directions, a two piece ring with a step gap may be used (Fig. 26, bottom). The joint mating surface of this ring is made perfectly flat and parallel to the side sealing face for better sealing.

Sealing in the radial direction is done by the inner member. The mating surface between these two rings, the inner and the outer, must be almost perfectly circular. For the most part, leakage in this ring is through the joint.

Pressure And Friction

For rings that must give good dynamic and static sealing, the sealing forces from hydraulic pressure must be sufficiently large in *both* radial and axial directions. This prevents hangup from friction on either the radial or the axial sealing surface.

PRESSURE BALANCING

Whenever there are heavy hydraulic pressures in a one-direction sealing ring, a good reduction in wear can be obtained by *pressure balancing*. However, balancing usually increases leakage. A circumferential groove is cut in the wear surface of the sealing member during manufacture, leaving a very thin dam which acts as a pressure reducer.

SPRING LOADING

For low pressure sealing, garter springs are sometimes used. It may also be helpful to use axial springing of some sort, such as the wave washer under the oil control ring in Fig. 24.

Maintenance of Split Ring Seals (Piston Rings)

Since engine piston rings are the prime example of split ring seals, most of our discussion will use them as an example. Almost all these maintenance problems apply to other split ring seals as well.

Engine piston rings seal by springing out or expanding against the fixed cylinder wall as the piston moves up and down. The rings also seal against the side of the ring groove in the piston. This is mostly due to pressures in the cylinder. The rings are usually made of cast iron and are often chrome plated on their sealing surface for longer life.

The main function of the piston ring is to prevent leakage of compressed fuel and air or expanding, burning gases around the piston into the engine crankcase.

Another function is to control the amount of oil for piston and ring lubrication and prevent excessive flow of oil past the piston.

To perform all these functions, close tolerances at the rings must be maintained.

Inferior oils do not maintain a good film of oil between the moving parts. Consequently, there is more metal to metal contact, and wear occurs. Then too, the thin oil film does not dissipate enough cylinder heat and is more likely to become carbonized. Lack of important additives may lead to ring scuffing and possible piston seizure. Premium oils are good insurance for trouble-free operation.

Fig. 27 — Plugged Oil Control Ring On Engine Piston

Failure to maintain proper operating temperature can also contribute to excessive ring wear or sticking (Fig. 27). Ring grooves in the piston may also wear, mostly on the bottom of the groove.

High operating temperatures, low coolant level, or obstructions in the cooling system can cause carbon to form in the ring area and make the rings stick.

If the operating temperature isn't high enough, the fuel-air mixture may condense in the engine cylinders. This condensate washes the lubricant from the cylinder walls, permitting metal to metal contact and wear. The condensate also mixes with the crankcase oil, further diluting it and reducing its lubrication capability.

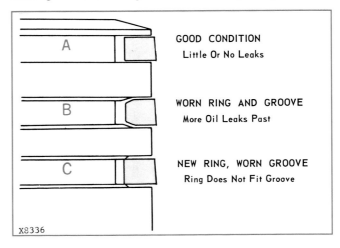

Fig. 28 — Piston Ring Grooves With
Taper Faced Rings (Schematic Veiw)

When piston rings stick, the condition is aggravated because of blowby of the hot gases past the piston. This causes the piston to run hotter, more carbon to form in the ring area, and often causes scoring due to burning the oil off the cylinder walls.

Keystone rings are often used in high speed diesel engines to help reduce sticking. Also inserted grooves help reduce wear (Fig. 29). Some rings are "plasma" coated to give them anti-scuff characteristics.

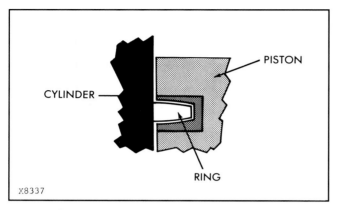

Fig. 29 — Keystone Ring With Inserted Groove

The purpose of the piston ring is to seal, not only between the ring and the cylinder wall but also between the ring and the piston.

Fig. 28 illustrates a piston with different types of ring groove conditions.

Diagram (A) illustrates the ring and groove in **new** condition. When the gases in the cylinder are compressed or expand during combustion, the pressure forces the ring to seat against both the ring groove and the cylinder wall. So, if ring and groove are in good condition, little or no leakage occurs.

Incidentally, when the piston moves down on the intake stroke, the piston ring seats against the top of the groove. As the ring descends, it scrapes oil from the cylinder wall. This oil lubricates and seals the ring.

As the piston ring wears, a wear pattern results as illustrated in (B), Fig. 28. When the ring wears, tension is reduced and the clearance in the groove is increased. As the piston moves down on the intake stroke, the increased clearance scrapes more oil from the cylinder walls and traps it behind the piston ring. When the piston moves up, this creates a "pumping" action which results in excessive oil consumption.

Diagram (C), Fig. 28 illustrates the condition when a **new** ring is installed in a **worn** groove. It is obvious that the ring cannot provide an adequate seal in the groove. This is the main reason why, when only new rings are

installed, the engine may still use too much oil.

When piston ring grooves become worn, install new pistons and rings. Do **not** attempt to recondition piston ring grooves, use oversized rings, or reuse old piston rings. Also inspect the condition of cylinder bore and keep parts clean during assembly.

The above methods cannot be used on all pistons. Always consult the engine Technical Manual for specific information.

NOTE: For more details on service of engine piston rings, see the FOS manual on "Engines."

INSTALLING RINGS

Take extra care when installing piston rings. Never expand them more than necessary to place them over the piston. Avoid twisting as this will distort them and prevent seating against the cylinder walls. Also inspect ring grooves to make sure they are free of nicks. If you install new rings on an old piston make sure that ring grooves are free of deposits.

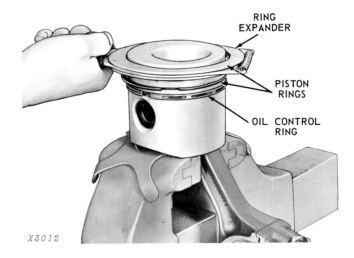

Fig. 30 — Installing Piston Rings Using A Ring Expander

It is recommended that a piston ring expander (Fig. 30) be used to install piston rings. This expander not only prevents twisting the ring but also assures that it is expanded evenly all around, preventing distortion.

IMPORTANT: The old practice of installing piston rings by hand is definitely wrong. This method almost always distorts the rings.

Many rings in current applications are taper faced. Make sure taper face rings are installed correctly (Fig. 28).

Always oil the piston and rings before installation.

Fig. 31 — Installing Piston In Cylinder

Many types of piston ring compressors for installing pistons in the cylinders are available. However, the type illustrated in Fig. 31 is by far the most satisfactory. It has a metal sleeve with the inside tapered so that the rings are gradually compressed and easily enter the cylinder. This avoids possible burring or distortion which can occur when other types of compressors are used.

Metallic ring seals are also used in transmissions, oiling shafts and on hydraulic piston heads to prevent oil leakage where nonmetallic seals would be inadequate because they would not be able to withstand the pressures.

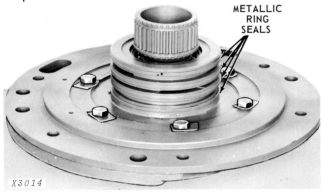

Fig. 32 — Metalic Seal Rings

Fig. 32 illustrates a typical installation of metallic sealing rings in a hydraulic transmission. The metal rings seal like engine piston rings and must be free to move in their grooves. Any burrs or sludge will restrict the movement and impair the sealing action.

Special tools should be used to remove and install these rings. Special ring expander pliers are often recommended. The rings themselves usually have hooked ends for blind assembly.

Be careful not to expand the ring more than necessary to install it.

Care must also be taken when installing the component that carries the sealing rings. Usually the opening is chamfered. But be careful. Avoid burring the rings.

CIRCUMFERENTIAL SEALS

Circumferential seals are high performance, low leakage, contact type seals for use in rotary applications.

They can accept high velocity rubbing at their primary sealing surfaces and offer a high sealing ability. They will also permit unlimited axial motion between stationary and rotating members.

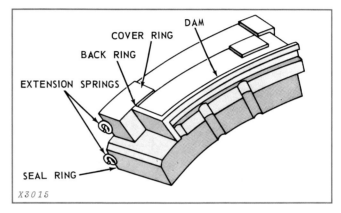

Fig. 33 — Circumferential Seal

These seals require smaller space than face seals and are lightweight.

Circumferential seals are quite similar to rod and piston seals, but are adapted to rotary motion. They are more complicated than regular split ring seals and always use springs for pressure in sealing.

Basic Features

The major parts of a circumferential seal are one or more relatively stationary, flexible, split or segmented rings (Fig.. 33) which mate with a rotating member. The rings are loosely keyed to the stationary member of rotation locks.

Circumferential seals prevent leakage in two directions.

Primary sealing takes place at the cylindrical surface where high speed relative motion occurs.

Maintenance Problems

Out of round or taper of the primary sealing surface will result in high leakage and high wear. Moreover, as it wears, the seal ring must move radially to maintain contact. The required movement may not initially be uniform around the periphery. This creates a problem in maintaining total contact of the seal ring with the runner throughout the life of the seal.

Heat and the resulting wear can also be a problem with circumferential seals. Carbon is widely used in the sealing ring for least wear. In some high-temperature uses, cooling oil is fed to these seals during operation.

FACE SEALS

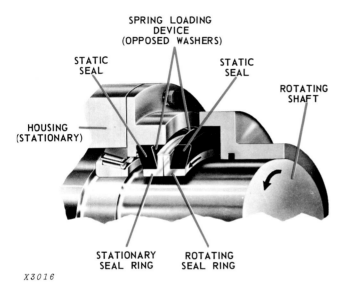

X3016

Fig. 34 — Face Seal (Axial Mechanical Seal)

The *face seal* or *axial mechanical seal* forms a running seal between flat, precision finished surfaces (Fig. 34).

Used for rotating shafts, the sealing surfaces usually are located in a plane at a right angle to the shaft. Forces which hold the rubbing faces in contact are parallel to the shaft.

These seals replace conventional stuffing boxes where a fluid must be contained in spite of a substantial pressure head.

All face seals have the following elements.

1. **Rotating seal ring**
2. **Stationary seal ring**
3. **Spring loading devices**
4. **Static seals**

The rotating and stationary seal ring are spring loaded together, and sealing takes place on their rubbing surfaces. The static seal prevents leakage past the juncture of the rotating seal ring and the shaft. See Fig. 34.

Since the rotating seal ring is stationary with respect to the turning shaft, sealing at their junction is done with gaskets, O-rings, V-rings, cups, etc.

GENERAL TYPES

The main advantage of an end face seal is its very *low leakage* rate.

The face seal also causes *little wear* on the sleeve or shaft it seals. Dynamic sealing is created on the seal faces in a plane vertical to the shaft and very little relative movement occurs between the seal head and the sleeve or shaft on which it functions.

In most instances, the face seal is automatically assembled into a box or set into the unit.

Standard end face seals have been used under pressure up to 3000 psi (20 685 kPa) in heavy duty process pumps. Special designs have been developed for pressures up to 10,000 psi (68 950 kPa). They allow for a determined gap between seal faces.

Another method of handling high pressure is the tandem seal arrangement. By breaking down pressures in respective chambers, each seal faces only the resulting pressure differentials. As many tandems as necessary may be added.

MAINTENANCE OF FACED SEALS

Lubrication of the seal face must be suitable. Coking, resulting from thermal cycling, can hamper seal movement.

Face materials which *run dry* because of equipment breakdown may fail prematurely. Some double seals with isolated liquid circulation avoid this hazard.

Abrasives can damage sealing faces if the seal is not properly lubricated and flushed during operation.

When installing new face seals, be sure to assemble them properly and locate them correctly using the recommended drivers.

METAL BELLOWS SEALS

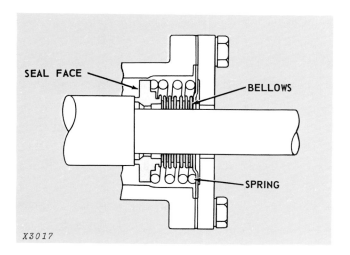

Fig. 35 — Metal Bellows Seal On Shaft

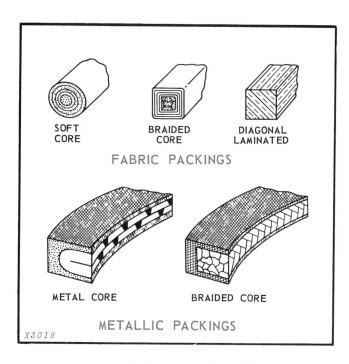

Fig. 36 — Typical Compression Packings

The main function of the *metal bellows* in face seals (Fig. 35) is to replace the static seal element.

Their major advantage is special applications where an *all metal* seal is necessary; generally for high temperatures or where the liquids or gases used would harm other seal materials.

In addition, the bellows can be used as a spring to produce the face loading needed for sealing contact between the seal faces. In some seals the bellows spring force is supplemented by a helical coil spring (Fig. 35). In others, the bellows provide all of the required spring force.

Maintenance

Metal bellows seals are built of metal for severe service. Properly installed and cared for, they will give long service life.

COMPRESSION PACKINGS

Compression packings create a seal when squeezed between the throat of a stuffing box and its gland.

Under these compressive forces, the packings flow outward to seal against the bore of the box and inward to seal against the moving shaft or rod.

Compression packings require frequent gland adjustment to compensate for wear.

DESIGN

Compression packings are usually made square or rectangular in cross section. There are three classes of packings: **fabric, metallic,** and **plastic.**

Several types of construction (Fig. 36) are used for fabric and metallic packings to provide resilience.

The packing ring joint is important. A diagonal cut gives good overlap at the joint, and makes it possible for gland and fluid pressure to create a tight seal. It also allows the ends to slip over each other if and when the ring expands. If multiple rings are used, the joints are staggered to prevent leakage through the end cuts.

OPERATION

A packing is like a bearing, it needs lubrication.

Usually the packing is snugged up by means of the gland. In operation, the fluid flows through the slight clearance between the moving parts and the packing, and acts as a lubricant for the packing.

If the gland is overtightened, *all* leakage is stopped, causing the packing to run dry and create frictional heat if the packing seals a rotating or reciprocating shaft.

For high temperatures or pressures, some means of external lubrication and cooling may be necessary.

SELECTION

The cross section of a packing is related to the size of the rod or shaft where it is used. Recommended packing sizes for various shaft sizes are listed below.

Recommended Packing Sizes for Various Shaft Diameters

Shaft Diameter Inches (mm)	Packing Size Inches (mm) in Cross Section
1/2 to 5/8 (13–16) 5/16 (7.9)	
11/16 to 1-1/2 (17.5–38) 3/8 (9.5)	
1-9/16 to 2 (40–51) 7/16 (11.1)	
2-1/16 to 2-1/2 (52–64) 1/2 (12.7)	
2-9/16 to 3 (65–76) 9/16 (14.3)	
3-1/16 to 4 (78–102) 5/8 (15.9)	

If an **undersized** packing is used, it must be compressed too much to fill the packing space. This reduces the amount of compression provided for future takeup.

On the other hand, an **oversized** packing must be beaten or rolled down so that it fits in the packing space. This overloads the packing and causes it to overheat.

MAINTENANCE

Packings should be installed carefully to avoid twisting or tearing. Lubricating them is usually recommended before installation. Some are also soaked in oil or other fluid.

As with most other seals, it is a good rule to **replace all packings which are disturbed during repair.** Always use the correct size of packing for replacement.

MOLDED PACKINGS

Molded packings are often called automatic, hydraulic, or mechanical packings.

These packings usually do not require gland adjustment after installation. The fluid being sealed supplies the pressure to seal the packings against the wearing surface. Molded packings are normally used for *dynamic* sealing.

There are two major types of molded packings:

• **Lip**
• **Squeeze**

LIP packings include flange, cup, U-cup, U-ring, and V-ring packings.

SQUEEZE types include O-rings and related forms, plus felt packings. All rely on interference built into the ring for effective sealing.

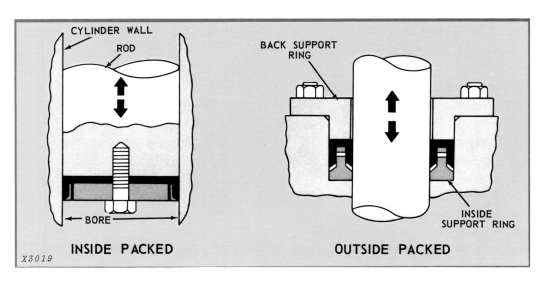

Fig. 37 — Inside And Outside Packed Installations

LIP-TYPE PACKINGS

Lip packings of the flange, cup, U-cup, U-ring, and V-ring configurations are used almost exclusively for dynamic applications.

Although rotary motions are found, the packings discussed here are used primarily for sealing reciprocating parts (Fig. 37).

Lip packings are usually made of leather, rubber, or synthetics (often rubber-fabric mixes).

General Design

SEAL LOADING: The principle of operation for all dynamic packings is the same, regardless of type of material. A molded packing, particularly the lip type, should be installed so that it can expand and contract freely. This action is referred to as *breathing.* A molded packing should not be used under high mechanical pressure, since this in effect transforms a self acting packing into a compression packing.

A lip-type packing is sealed by the force of the pressure being sealed, which causes the packing to expand. Friction force is produced in proportion to the pressure on the pressure stroke. If the packing is allowed to breathe, the friction on the return stroke is negligible.

Preloading is essential to ensure a tight fit for packings at low pressures. A lip-type packing, molded with a slight flare, is preloaded at the lips upon installation.

BACK SUPPORT RING: Under pressure, all packings follow the shape and contour of the back supporting components. Since the greatest wear on a packing occurs at the shoulder, proper support of the packing at its shoulder is the most vital factor in the design.

If excessive clearance exists between the cylinder wall or the shaft and the ring, operating pressure will extrude the shoulder of the packing into the clearance.

Examples of typical inside and outside dynamic packing installations are shown in Fig. 37. Recommended clearances are based on the type of material.

INSIDE SUPPORT RING: An inside support ring, if made from metal, should not have a tight fit, but should have clearance. The corners must be chamfered to prevent cutting the inside shoulder of the packing.

If no clearance is provided, the packing will swell and bind. With clearance, the packing is free to breathe.

METALS AND FINISH: The type of metal and surface finish over which the packing operates is critical. A hard, close grained material that will take and hold a fine finish produces the least packing friction and wear. Rods exposed to the atmosphere are normally chrome plated. In a dirty environment, the packing and other parts are usually protected by a rod scraper or wiper seal.

BEARING AREA: Packings are designed to function as a seal and should not be subjected to any bearing loads.

CYLINDERS AND RODS: Good packing performance cannot be expected from cylinders or rods which are not concentric.

SYSTEMS: A clean lubricating system is vital. Filters and rod scrapers will guard against contamination. A filter in a line will trap foreign particles, such as metal chips and dirt, and help ensure long packing life with trouble free service.

OPERATING CONDITIONS: The packing used depends on whether it is for inside packed or outside packed installations. Some styles can only be used for one or the other.

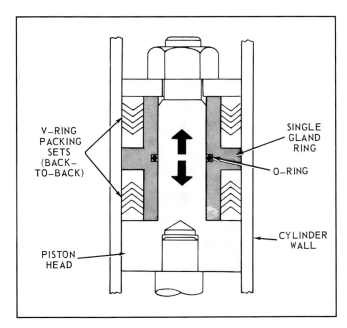

Fig. 38 — Inside Packed Installation

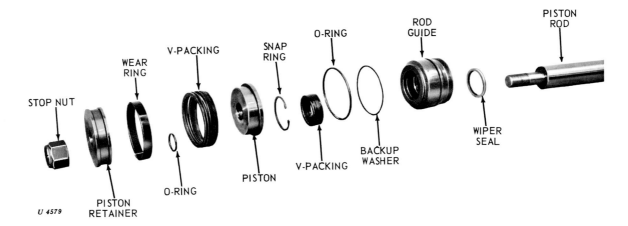

Fig. 39 — V-Packings On Double Acting Piston Installation

In an **inside** packed installation (Fig. 37, left), the packing is on the end of the rod or plunger and moves with the rod. The following styles can be used: cup, U-ring, and V-ring.

In an **outside** packed installation (Fig. 37, right), the packing is in a stationary gland and the rod moves through the packing. Here the flange, U-ring, and V-ring styles are used.

No one packing style is superior to all others. Of the four styles, the cup and flange are considered unbalanced, while the U-ring and V-ring are considered balanced. A balanced packing seals on both the O.D. and I.D. and the pressure is equalized on the side walls of the gland or recess.

V-Ring Packings

This is one of the most popular of all the lip types, and is used for low or high pressure applications. It can be installed on a piston (inside packed) or in a gland (outside packed.) However, it is primarily used for outside packed, while the U-ring is used primarily for inside packed. V-ring packings are shown in Figs. 38 and 39.

V-ring packings are installed in sets, each set consisting of a number of V-rings (pressure rings) and a male and female adapter. The number of rings to a set depends upon the pressure and packing material.

V-ring sets are often split on an angle for easy installation. When split sets are used, the joints should be staggered on successive packings, like piston rings. Whenever possible, endless (solid) rings are used because of their better sealing and longer life.

PACKING APPLICATION: Homogeneous and most fabric V-ring packings are generally molded with the lip at a slight angle. When the packings are installed, this flare automatically preloads the packings, and only slight initial adjustment of the gland ring is necessary.

Leather V-rings do not have this flare and require slightly more initial adjustment to provide the preload.

Regardless of the material, packing sets should not be tightened excessively, but should be allowed to function as a mechanical packing. Due to their shape, they are sensitive to pressure changes. Sets of packings should not be loose in a gland or they may become cocked. Furthermore, a loose set will cause a pumping action that results in leakage during fast reversals.

Fig. 38 shows a design for a double acting cylinder with only one gland ring with two sets of packings. Another double acting piston with V-packings is shown in Fig. 39.

SPRING LOADING: Generally, V-ring packings are not spring loaded, since the set can be preloaded with an adjustable gland or without adjustment. However, some sets of V-ring packings are loaded with metal springs to eliminate manual tightening of the gland ring.

Cup Packings

Cup packings are widely used for hydraulic and pneumatic service at low and high pressures. They are usually made of leather, homogeneous, or fabricated synthetic rubber.

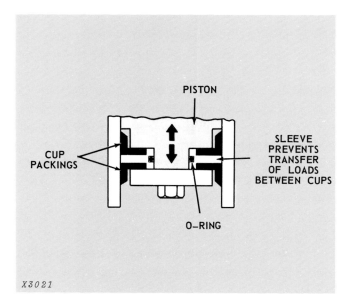

X3021

Fig. 40 — Cup Packings For Double-Acting Unit

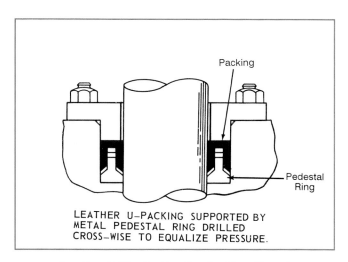

Fig. 41 — U-Ring Packing (Leather Type Shown)

Cups having only a single lip, as shown in Fig. 40, are classified as unbalanced packings.

Excessive tightening of the inside follower to the back follower crushes the bottom of the cup and hinders sealing.

At extremely low pressures, a cup may seal at the lip, but at higher pressures, the proper seal is not made until the heel diameter has expanded to the diameter of the cylinder wall. This causes wear to take place at the heel or shoulder of the packing. Very little wear, if any, occurs at the lip.

Resilient cup packings are preferred over hard cups except for very high pressures.

For double acting units, a boss design is often used to prevent the transfer of load from one cup to the other (Fig. 40). The sleeve should have a close fit in the cylinder to prevent extrusion.

The lips or edges of cups are usually beveled from 15° to a maximum of 30° from the horizontal.

Cup Packing Expanders: Constant pressures, if reasonably high, will not usually require an expander for the packing. However, low and variable pressures often require expanders to exert the proper mechanical pressure on the lip of the packing. This maintains a slight preload on the cylinder or plunger.

U-Ring Packings

This packing in leather is sometimes called a "double cup;" in homogeneous, a U-cup. For both types, the correct name is *U-ring packing* (Fig. 41).

They are balanced packings which seal on both the O.D. and I.D. They have low friction and are not stacked in nested sets.

The leather U-ring is more common than the fabricated, and the method of support and the configuration varies between the two.

LEATHER TYPES: The gland and back supporting ring for leather U-ring packings are important (Fig. 41). The packing must be properly supported to keep the lips from collapsing. Leather U-ring packings are made with straight sidewalls. The recess of the U-ring is filled with flax, hemp, rubber, hard tucks, or a metal pedestal ring as shown.

Clearance is required between each side of the pedestal ring and inside walls of the packing (Fig. 41). This clearance allows for swell and eliminates binding.

FABRICATED TYPES: Unlike the leather U-ring packing, the fabricated type is molded with the bottom thicker than the side walls and the side walls molded on a flare. This flare preloads the packing upon installation, and does not require a filler or any other support in the recess of the packing.

Except for the above difference, the fabricated type is similar in operation to the leather type.

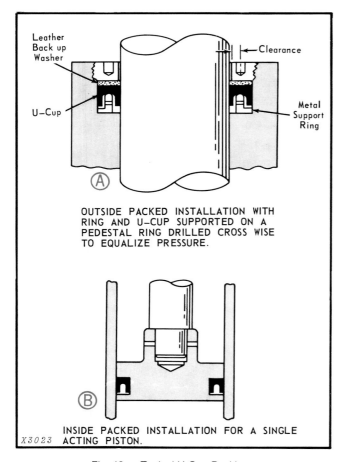

Fig. 42 — Typical U-Cup Packings

U-Cup Packings

This style of balanced packing is primarily made of homogeneous synthetic rubber, and is easily installed. It has a low friction and is primarily used for pressures under 1500 psi (10 340 kPa), although some designs are made for 5000 psi (34 475 kPa) and higher.

Typical U-cup packing designs are shown in Fig. 42.

A metal support ring is often used for *outside* packed installations (Fig. 42, A).

For *inside* packed installations, the metal support ring is seldom used. Fig. 42 (B) shows an inside packed installation for a single acting piston with a U-cup stretched over the piston and snapped into place.

For a double acting piston, an additional groove is machined and two U-cups are installed. Two U-cups are never used in the same groove. A two-piece piston is sometimes used to avoid groove cutting and consequent stretching of the U-cup during installation.

The U-cup replaces the cup packing in many installations; often it is simpler and cheaper.

Flange Packings

The flange packing, sometimes called "hat," is the least popular of all the lip-type packings.

Primarily made of leather, it is also produced in fabricated and homogeneous materials. It is unbalanced, with sealing on the I.D. only, and is used for low pressure, outside packed installations where a V-ring or U-ring cannot be used due to lack of space. It is used for both reciprocating and rotary motion.

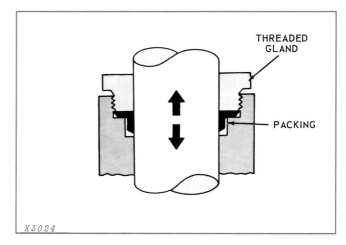

Fig. 43 — Flange Packings

In most installations, the flange packing is installed using a threaded gland (Fig. 43).

The packing is mechanically operated by the actuating pressure on the I.D. side wall and relies upon mechanical compression of the base for sealing the O.D. Hence, the gland ring must be drawn up tight enough to seal for the maximum operating pressure.

Clearance is required between the side wall of the packing and the cylinder body to allow for swell. There also must be clearance between the beveled lip of the packing and the bottom of the gland.

Standards have been established only for leather flange packings up to 5 1/2-inch (140 mm). For larger rods, V-ring or U-ring packings are used.

SQUEEZE PACKINGS

Molded squeeze packings are made in a variety of sizes and shapes, but nearly all of them offer these advantages: low initial cost; compact, easy to install; efficient; no adjustment; tolerate wide ranges of pressure, temperature, and fluids; seal in both directions; and relatively low friction.

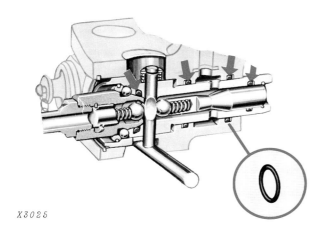

X3025

Fig. 44 — Squeeze Packings (O-Rings Shown)

Squeeze packings are generally fitted to rectangular grooves machined in hydraulic or pneumatic mechanisms.

Since the rubber **O-ring** is the most common form of squeeze packing, and its application is similar to other types, this section is mainly about them.

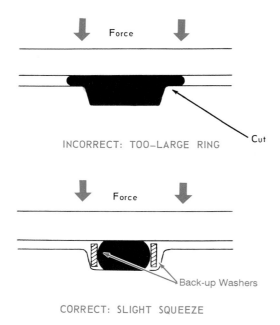

Fig. 45 — O-Ring Squeeze

Operation

O-rings work because they are *squeezed* when installed (Fig. 45). But it is the pressure from the confined fluid that produces the final deformation which causes the elastic O-ring to seal.

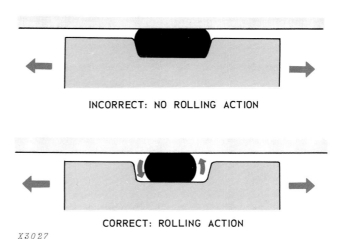

X3027

Fig. 46 — O-Ring Rolling Action

There are three uses of dynamic O-rings:

1. **Reciprocating,** where the sealing action is that of a piston ring or seal around a piston rod.

2. **Oscillating,** where the seal rotates back and forth through a limited number of degrees or several complete turns.

3. **Rotating,** where a shaft turns inside the I.D. of the O-ring.

Grooves for *reciprocating* motion are from 25 to 50 percent wider than the diameter of the O-ring cross section. The O-ring then seals by rolling and sliding to the side of the groove away from the pressure. When the pressure forces reverse, the ring rolls and slides to the other side of the groove (Fig. 46).

O-ring grooves are normally chamfered to avoid cutting the O-rings.

Back-up washers made from plastics are often used to stabilize the O-ring motion and prevent damage to the O-ring. They may be used on one side or both sides of the O-ring (Fig. 45).

Applications

Squeeze packings are economical and easy to install. One common application of an O-ring is as a piston seal in a hydraulic actuating cylinder. Another uses the O-ring as a valve seat or as a valve plug packing (Fig. 44).

O-rings are not used: (1) in very dirty or abrasive conditions; (2) where lubrication is sparse; (3) at very high shaft speeds.

Backup rings are sometimes used with O-rings to prevent "nibbling" and other damage in operation.

Maintenance Of O-Rings

O-rings can be easily damaged by cutting or nicks from sharp objects. They can also be damaged by heat, wrong fluids, lack of lubrication, and improper installation. See Fig. 47 below.

INSTALLATION OF O-RINGS

1. Be sure the new O-ring is compatible with the hydraulic fluid. Otherwise, the O-ring may corrode, crack, or swell in operation.

2. Clean the dirt and grit off the entire area before installing O-rings.

3. Inspect O-ring grooves before installing rings. Remove any sharp edges, nicks, or burrs with a fine abrasive stone. Then reclean the area to remove any metal particles.

4. Inspect the shaft or spool (if used). Sharp edges or splines can cut O-rings. Remove any nicks or burrs with a *fine* abrasive stone. Then polish with a *fine* abrasive cloth. Reclean the area to remove any metal particles.

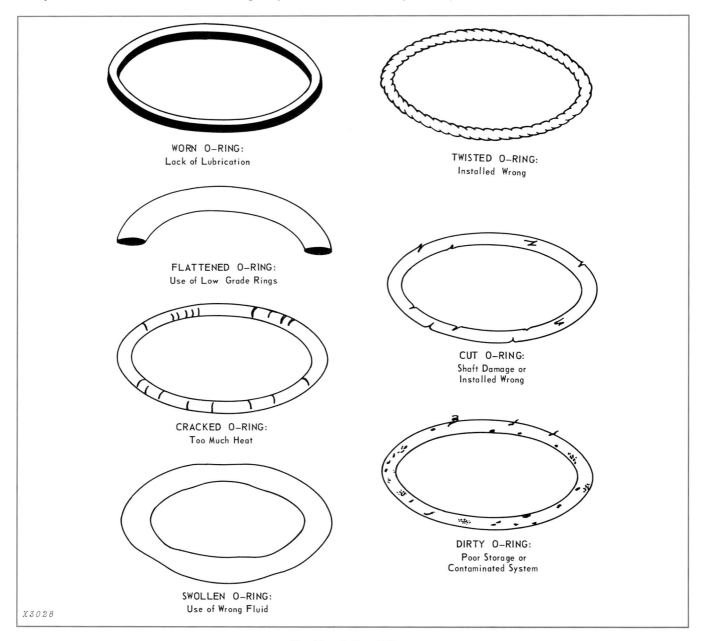

WORN O-RING:
Lack of Lubrication

TWISTED O-RING:
Installed Wrong

FLATTENED O-RING:
Use of Low Grade Rings

CUT O-RING:
Shaft Damage or
Installed Wrong

CRACKED O-RING:
Too Much Heat

DIRTY O-RING:
Poor Storage or
Contaminated System

SWOLLEN O-RING:
Use of Wrong Fluid

X3028

Fig. 47 — O-Ring Failures

5. Lubricate the O-ring before installing it. Use the same fluid used in the system. Also moisten the groove and shaft with the system fluid.

NOTE: Some O-rings are identified by paint marks that tell the O-ring has a special use; will resist certain temperatures of fluids. Be sure to match new O-rings with old ones.

6. Protect the O-ring from sharp edges, and be careful not to stretch it more than necessary while installing it.

7. Align the parts accurately before mating to avoid twisting the ring.

8. Check to see that the O-ring is the correct size to give only a slight "squeeze" in the installed position (Fig. 46). In dynamic use, the O-ring should be able to roll in its groove.

9. Always install the proper back-up washers, if required.

IMPORTANT: When installing spool valves, and similar parts, be especially careful of O-rings. The sharp edges of the lands can cut the O-rings unless you are very careful.

CHECKING O-RINGS AFTER INSTALLATION

Dynamic O-rings should be cycled back and forth through their normal pattern of travel several times to allow the ring to rotate and assume a neutral position.

All dynamic rings should pass a very small amount of fluid when rotating to lubricate between the ring and shaft.

NOTE: Static O-rings are covered later in this section.

FELT RADIAL PACKINGS

Felt is a fabric made of wool fibers or a combination of wool and animal, vegetable, and synthetic fibers.

Uses

Here are some reasons why felt is used as a sealing material:

1. *Oil Wicking:* Capillary properties of a felt seal assure lubrication after long idle periods.

2. *Oil Absorption:* About 78% of the volume of a felt seal serves as oil storage.

3. *Filtration:* Wool felt seals, in a dry state, are 99 to 100 percent effective as filters in removing small particles. When saturated in lubricants, even more particles are trapped and retained.

4. *Resiliency:* Felt seals maintain constant sealing pressure in spite of wear, and play, minor misalignment, or out of roundness of metal components.

5. *Polishing Action:* Felt seals trap abrasive particles, which then penetrate into the felt where they can't score the part.

Types

Felt seals are manufactured in two general types:

• *Plain*

• *Laminated*

To meet special requirements, either style can be impregnated with one of several materials.

Plain felt seals are precision cut washers fabricated from standard grades of felt. They are usually presaturated with lubricants of slightly greater viscosity than that used in the bearing. Plain felt seals afford positive bearing protection and provide a reservoir for lubricant storage. If run dry, they protect and polish rather than score a shaft. They seldom fail from aging or disintegration. These seals are not used with oils of low viscosity or to retain pressurized lubricants. For these applications, laminated or impregnated felt should be considered.

Laminated felt seals are made by combining one or more grades of felt with layers of impervious elastic. This permits use of two or more felt grades or densities in one seal; one grade for oil retention and a second for dust exclusion. The impervious layers of oil resistant elastic also stop leakage of low viscosity fluids through the felt.

Maintenance

Replace felt seals when a system is overhauled.

Be sure the seal is lubricated and avoid stretching it too much when installing it over a shaft.

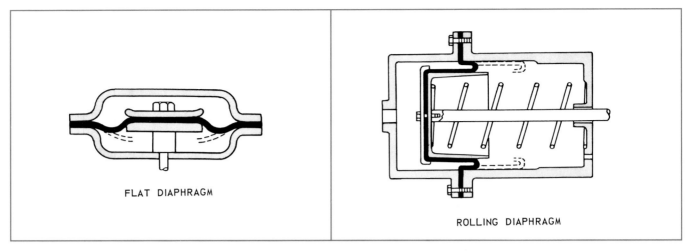

FLAT DIAPHRAGM

ROLLING DIAPHRAGM

Fig. 48 — Diaphragm Seals — Two Types

DIAPHRAGM SEALS

A **diaphragm** is a dividing membrane which spans the gap between a moving and stationary member. As a seal, it prevents interchange of a fluid or contaminant between two separated areas or chambers.

Diaphragms can be divided into three types:

1. *One type acts solely as a **separating membrane** where no pressure differential exists between the chambers.*

2. *A **static diaphragm** acts as a separator between two fluids and is subject to little or no motion.*

3. *A **dynamic diaphragm** acts as a sealing device between stationary and moving members, and usually transmits a force or pressure.*

The diaphragm used as a *dynamic* sealing membrane and force transmitter is the most common type and the main part of this section is devoted to it. This diaphragm acts in the same way as a sliding contact packing, except there is no leakage.

There are two kinds of dynamic diaphragms:

• **Flat**

• **Rolling**

A FLAT diaphragm (Fig. 48) is one which has no convolutions, or convolutions which are less than 180°.

A ROLLING diaphragm (Fig. 48) has a full 180° convolution; during its cycle the diaphragm rolls off the wall of the cylinder or piston onto the other cylinder or piston wall.

FLAT AND CONVOLUTED DIAPHRAGMS

"Flat" diaphragms are not necessarily flat. Even when a diaphragm is made flat, convolutions are formed by fluid pressure during the working cycle. Also, diaphragms which are made flat are distorted when there is relative movement of the diaphragm assembly (Fig. 48, left). This distortion requires an elongation of the fabric cord length. For this reason, flat diaphragms are generally made of a flexible material. This allows the diaphragm to flex rather than stretch.

Diaphragms are usually made of fabric, impregnated with an elastic. Most combinations can withstand flexing indefinitely, but excessive stretching can cause fatigue failures.

Diaphragms are attached to the housings at the edges, while the center is usually held with plates. A *push rod*, attached to the center plates and guided by the housing, transfers motion to the diaphragm, or is moved by pressure behind the diaphragm, See Fig. 48, left.

Springs are usually used in conjunction with diaphragms to provide a return or a reference force for pressure regulation.

For each application, there is a correct "squeeze" or precompression which provides adequate sealing without undue distortion of shape.

ROLLING DIAPHRAGMS

Rolling diaphragms are used as if they were long travel bellows, or radially sealed hydraulic pistons with zero leakage. See Fig. 48, right.

Operation

The rolling diaphragm has a full 180° roll in operation. When pressure increases in the loading chamber, the piston moves down, causing the diaphragm to roll off the piston sidewall and onto the cylinder wall with a frictionless action (see Fig. 48, right). Almost the entire pressure load is supported by the piston head, and only a small load is felt by the diaphragm. Contained pressure holds the rolling diaphragm against the walls.

Rolling diaphragms can be used in many medium and high pressure applications.

However, rolling diaphragms cannot be used where pressure reverses and pressure in the low pressure side exceeds pressure on the high pressure side by more than 1/2 psi (3.4 kPa). This pressure reversal may cause sidewall distortion, overstressing, and scuffing.

STATIC SEALS

This section describes all the **static** seals listed in the box on page 44.

NONMETALLIC GASKETS

NONMETALLIC GASKET

X3030

Fig. 49 — Nonmetallic Gasket In Low-Pressure Application

Gaskets are **static** seals. A gasket is any device which maintains a barrier against the transfer of fluids across mating surfaces of a mechanical assembly *when the surfaces do not move relative to each other.*

Joint and gasket design must be considered together. A joint is only as good as its gasket, and the gasket may succeed or fail according to whether the joint makes the best use of the gasket material. Therefore, joint components must be thought of as a unit, or system, for effecting a seal. Otherwise, the end result may be a leaky joint.

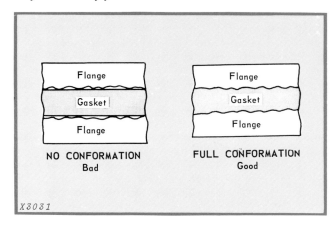

Flange
Gasket
Flange
NO CONFORMATION
Bad

Flange
Gasket
Flange
FULL CONFORMATION
Good

X3031

Fig. 50 — Conformation Of Gasket To Flange

A prime factor in any seal or gasket is *minimum sealing stress.* This is the minimum stress necessary to make the seal material conform to the imperfections on the flange face (Fig. 50) and, where necessary, to close the material's porous structure so leakage does not occur.

JOINT AND GASKET DESIGN

The minimum stress required for sealing a gasket depends on:

• **Gasket material**

• **Internal pressure**

• **Sealed fluid**

• **Width to thickness ratio of gasket**

• **Gasket shape factor**

• **Flange finish**

Gasket Material. Minimum sealing depends greatly on the type of material used in the gasket. Common materials are: cork, rubber, plastic, and sand paper, or combinations of these.

Internal Pressure. Internal pressure tends to force the sealed fluid through the seal assembly; thus, it has a vital effect on how much initial flange pressure is necessary to seal the joint.

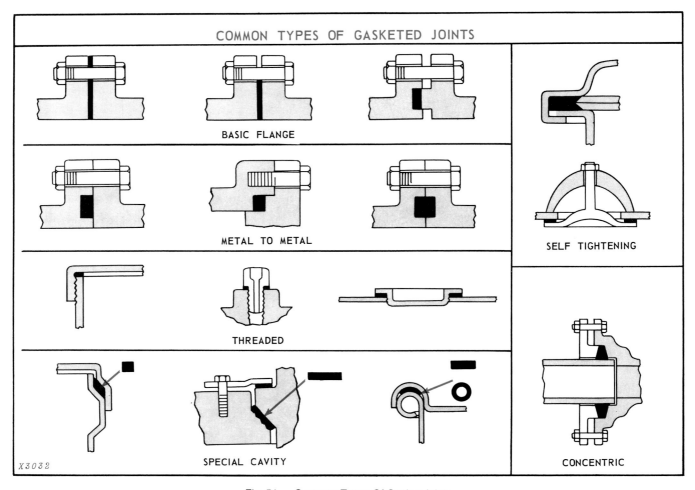

COMMON TYPES OF GASKETED JOINTS

BASIC FLANGE

METAL TO METAL

SELF TIGHTENING

THREADED

SPECIAL CAVITY

CONCENTRIC

X3032

Fig. 51 — Common Types Of Gasket Joints

Sealed Fluid. Except for certain materials like cork and rubber, the sealed fluid affects the required minimum sealing stress. Most important here is the type and viscosity of the fluid.

Width/Thickness Ratio. Minimum sealing stress decreases as the ratio of sealing width of a gasket to its uncompressed thickness increases. This is especially significant in gaskets which have bolt holes. In such cases, the holes must not be too close to the inner edge of the gasket.

Shape Factor. A satisfactory seal requires minimum sealing stress *and* uniform gasket compression. The most important factor for meeting both requirements at once is the shape of the gasket.

Shape factor has no effect on the load compression of cork composition, but a significant effect on rubber. The more compressible the material, the less effect shape has on its load compression, since loading does not cause as much lateral expansion.

Flange Finish. Flange surface finish is not an important factor with compressible gaskets because they conform readily to minor irregularities. However, harder materials may not deform enough to seal rough finishes.

In general, a gasket must be impermeable to fluids, conform to imperfections when compressed, seal through variable temperatures and pressures, and resist deterioration.

Common types of gasketed joints are shown in Fig. 51.

MAINTENANCE PROBLEMS WITH GASKETS

Some common problems with gaskets and gasketed joints.

Joint Distortion

Distortion of gasketed joints usually causes a non-uniform pressure distribution over the gasket. Joint distortion exists whenever the flange faces are not

Six types of joint distortions can occur. The first three: *flange bowing, bolt hole distortion,* and *cocking* are caused by the effect of bolt loads on a flexible flange. The fourth is caused by *flange surface roughness.* The last two are caused by *excessive internal pressure.* Distortions normally occur in combinations.

Flange Distortion: This can cause leakage at points where compression is too low for sealing. It can also cause gasket failure by extrusion or rupture where flange pressure is concentrated.

Flange Bowing: Even in relatively rigid flanges, bolt loads cause a certain amount of bending in the flange between bolt centers. Flange bowing almost invariably causes leakage through the low compression regions between bolt centers. In field assembly, a full and improper torquing of all bolts will help prevent flange bowing (see "Engine Cylinder Head Gaskets," later).

Bolt Hole Distortion: Such distortion occurs around the bolt hole in a flange, most often in sheet metal flanges. It normally is a bowing effect. Causes are overtightening screws or not using proper washers.

Flange Face Roughness: Surface roughness can cause leakage if the gasket is hard and compressive loads are inadequate because of bowing or some other factor. Leakage caused by surface roughness can usually be reduced by the proper seal, the right torque, and/or gasket cement.

Distortion From Internal Pressure: Internal pressure adds to the loading on bolts and flanges. Excessive internal pressure can cause bolt elongation, add to flange bowing, and cause gasket blowout. In rigid joints, internal pressure normally causes leakage or blowout. In assemblies where the flanges distort, pressure causes the flange to lift off the gasket, resulting in leakage.
In the field, these gasket leaks or blowouts can usually be prevented by proper torquing of bolts holding the gasket.

Torque Loss: Just as important as creating the initial seal, is maintaining it throughout the life of the assembly. As soon as bolts have been tightened, various factors go to work to loosen or destroy the seal.
Bolt torques or flange pressures begin to drop because the gasket takes a "set." Mechanical vibrations may also contribute, and temperatures may rise. And, the sealed fluid begins to interact with the gasket material.

Watch several factors to help reduce torque loss in the field: *initial flange pressure and bolt length, bolt diameter, temperature, time, retorquing, gasket material and thickness, and sealing surface conditions.*

Initial Flange Pressure: Because initial flange pressure interacts with bolt length, the two are discussed together. In a gasketed joint, initial flange pressure is that attained immediately after the bolts are tightened; bolt length is taken as effective bolt length which includes the flange, washer, and compressed thickness of the gasket.

At low initial torque, bolt length has no effect. As initial torque increases, however, bolt length begins to have a measurable effect.

For this reason, bolts are tightened to their maximum elongation *below* the yield point.

Bolt Diameter: The smaller the diameter, the greater the bolt elongation for a given torque. As long as yield tension is not reached, this gives better torque retention. Also, the higher the torque, the more equally the total load is distributed among the bolts.

Temperature and Time: Torque loss increases with temperature. Time also works against holding the torque. Torque loss is most rapid immediately after tightening, with most loss occurring within 18 hours. Beyond that point, torque loss proceeds indefinitely, but very slowly.

Retorquing: To guard against torque loss during service, retorquing is often helpful. This is done by tightening the assembly, warming it up at a constant temperature for a given time, and then retorquing to the initial torque. This reduces torque loss significantly (see "Engine Cylinder Head Gaskets," later).

Gasket: The gasket material has a considerable effect on torque retention. Various materials are rated in the following order: cellulose fiber rubber, cork and rubber, cork composition, rubber.

Uncompressed gasket thickness also has a marked effect. Normally, the thinner the gasket, the better the torque retention up to the point where sealing is no longer adequate.

Sealing Surfaces: Any coating on gasket or flange sealing surfaces contributes to torque loss. This includes lubricants, uncured adhesives, or contaminants. Anything more than a thin film causes some torque loss.

Temperature Effects

In any seal, temperature always interacts with other factors. In addition to torque loss, temperature affects relative motion of the flanges, as well as chemical interaction between the sealed fluid and gasket material.

As temperature rises beyond the temperature the joint was tightened at, seal efficiency usually improves. This is probably due to softening of the gasket, which then "flows" into surface irregularities in the flange.

However, prolonged exposure to high temperatures causes hardening of the gasket material. Normally, this does not harm the seal because it occurs after the seal has been created. However, above the normal temperature limits the gasket begins to deteriorate.

STATIC O-RINGS (NONMETALLIC)

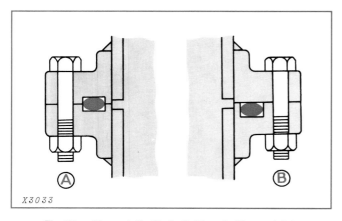

Fig. 52 — Nonmetallic Static O-Rings In Flange Joints
(Two Types Of Rectangular Grooves Shown)

Static O-ring seals are classified as gasket type seals but are in a class by themselves in terms of design and usage. See Fig. 52.

Static O-rings are generally easier to design into a unit than dynamic ones. Wider tolerances and rougher surface finishes are allowed on metal mating members. The amount of squeeze applied to the O-ring cross section can also be increased.

Static O-rings may shred or become nibbled or braded. Hence, contamination of filters and clogging of pumps are possible. No periodic tightening of nuts is necessary to maintain a tight joint. Once an O-ring is installed, operating pressure keeps it sealed against the metal clearance areas.

A rectangular groove is the most common one used for O-rings used as flange gaskets. The groove is machined one-half in the face plate and one-half in the flange (Fig. 52 A), or the entire groove is cut in one member (Fig. 52 B).

METALLIC GASKETS

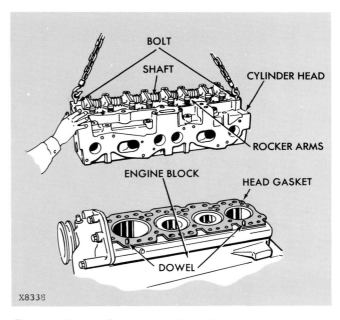

Fig. 53 — Metallic (Combination) Gasket On Engine Cylnder Head

Gaskets are softer than their mating surfaces and seal by conforming to the surface irregularities. However, two major kinds of gaskets are needed:

• **Nonmetallic gaskets—softer, for lower pressures**

• **Metallic gaskets—harder, for higher pressures and temperatures**

We covered nonmetallic gaskets in the preceding section. Now we will give details on metallic and combination gaskets.

GENERAL TYPES

Choice of a gasket material depends on operating conditions, mechanical features of the flanged assembly, and gasket characteristics.

In general, operating conditions govern choice of gasket **materials,** while features of the flange decide selection of the **type** of gasket.

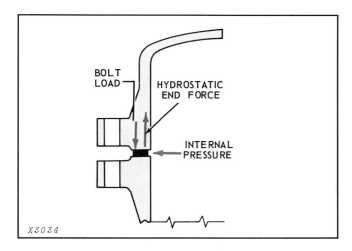

Fig. 54 — Forces Acting On Gasket Joint

How Metal Gaskets Are Selected

Pressure: Pressure of the confined fluid has little or no direct effect upon the selection of most types of metal gaskets. One exception is corrugated gaskets, which are generally limited to pressures less than 1000 psi (6895 kPa).

Temperature: Operating temperature is important in selection of the gasket type and material. The limitation of semimetallic gaskets is controlled by the non-metallic filler.

Joint clamping: Bolting must provide enough force to seat the gasket and to prevent flange separation due to the confined fluid or gas pressure (Fig. 54).

Joint rigidity: This must be considered because metallic gaskets usually take a permanent set when compressed in an assembly. Unlike non-metallic gaskets, they have little or no recovery to compensate for contact face separation. That is why joints for metal gaskets must be rigid to assure a minimum of bending during the initial bolting and when the assembly is operating. Flanges using full-face gaskets must be thick enough to prevent bowing between adjacent bolts.

Surface finish: This is important in that each type of metallic gasket performs best when the flange contact faces have a specific surface finish.

A problem with some high pressure cylinder head gaskets is that the metal finish can be *too smooth*, allowing the gasket to be pushed out by the pressure.

Corrosion: There are several different kinds of corrosion, and metallic gaskets are susceptible to most of them. Chemical resistance must often be built into the gasket material.

Seating Stress

Gasket sealing is done by flow of the gasket material into the imperfections on the flange facings. The amount of force per unit of gasket area required to completely flow the gasket is known as the *yield* or *seating stress*. This stress varies with gasket type, material, and flange surface finish.

Gasket Types

Metallic gaskets (Fig. 55) fall into several basic groups:

• *Corrugated or embossed, thin metal*

• *Metal jacketed, soft filler*

• *Spiral wound*

• *Plain or machined flat metal*

• *Round cross section, solid metal*

• *Heavy cross section, solid metal*

• *Light cross section, pressure actuated*

Corrugated: This type (Fig. 55) consists of thin metal, corrugated or with embossed concentric rings. They are used plain, coated at the time of installation with gasket compound, or with cord cemented in the corrugations.

Corrugated gaskets are essentially a line contact seal. Multiple, concentric corrugations provide a labyrinth effect, along with mechanical support for the gasket compound or cord inserts, if they are used. Corrugations provide some resilience, depending on their pitch and depth, and the type and thickness of metal used.

Metal Jacketed, Soft Filler: A soft compressible filler is partially or wholly encased in a metal jacket (Fig. 55). These gaskets are more compressible than corrugated types. They offer better compensation for flange irregularities when higher pressures are to be sealed.

TYPES OF METALLIC GASKETS

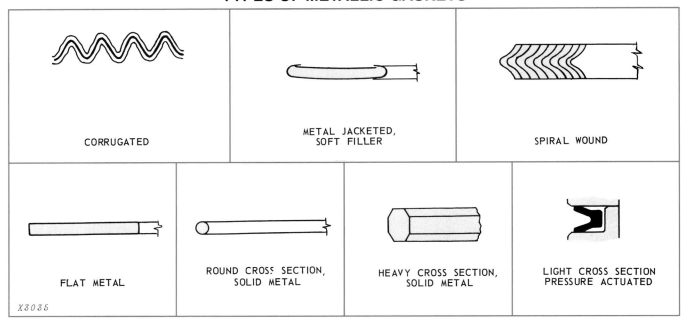

Fig. 55 — Types Of Metallic Gaskets

The primary seal against leakage is the inner metal lap, where the gasket is thickest when compressed. This area "cold flows," completing the seal. The entire inner lap must be under compression. The outer lap (if used) provides a secondary seal between flange faces when compressed.

These gaskets are used for noncircular as well as circular applications, and require 20 to 30% compression. They are widely used for engine cylinder heads (see later in this section).

Because of limited resilience, these gaskets are used only where the elasticity of the bolts or other factors can compensate for joint relaxation. They are not used in joints requiring close maintenance of the compressed thickness.

Standard filler used to be asbestos millboard. But now other filler material is used. Compressed filler sheet packing is used where higher strength is necessary. Metallic fillers are used when service temperatures exceed 900° F. (485° C).

Spiral Wound Gaskets: This type (Fig. 55) consists of V-shaped, preformed plies of metal, wound up in a spiral with a soft separation. The V-shape gives unique spring like characteristics.

Sealing action results from the flow of the metal and soft filler plies when the gasket is compressed the recommended amount. Both inner and outer metal to metal plies must be under compression. Spiral wound gaskets are particularly suited for assemblies subject to extremes in joint relaxation, temperature or pressure cycling, shock, or vibration.

Flat Metal Gaskets: These are gaskets that are relatively thin compared to their widths. They can be used as cut from sheet metal, or machined. Plain metal, washer shaped gaskets also perform satisfactorily in applications over a wide temperature range.

All types seal by flow of the gasket surface caused by brute force compressive loads. Loads actually must exceed the tensile strength of the gasket metal on the gasket contact area. Therefore, surface finish of both flange and gasket is very important.

Round Cross Section, Solid Metal: These gaskets are generally made from round wire, cut to the length of the gasket circumference, then formed into a circle and welded. They provide positive, gas tight seals at relatively low flange pressures. Since only line contact occurs, they have high local seating stress at low bolt loads. The contact faces increase in width as the gasket is compressed, effectively flowing into flange faces.

Solid, Metal, Heavy Cross Section: These are usually rectangular or triangular shaped gaskets, machined in heavy cross sections from solid metal. They are used for high pressure and high temperature service where operating conditions require special joint designs.

The gaskets usually seal by line contact or wedging action which causes surface flow. Some of them are pressure actuated; i.e., the higher the pressure, the tighter the joint.

Light Cross Section, Pressure Actuated: These metallic gaskets are used in place of elastic O-rings in sealing extremes such as vacuums and very high temperatures or pressures. They can also be used when standard metal gaskets are ruled out because great flange loads are required.

STATIC O-RINGS (METALLIC)

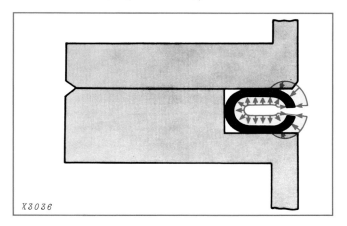

X3036

Fig. 56 — Typical Metallic O-Ring In Static Use

For a static seal that will meet extreme conditions of temperature, pressure, and corrosion with low cost and simplicity, the metallic O-ring is widely used. (Fig. 56).

Metallic O-rings are made of metal tubing which is formed into circular or other required shapes; the two ends are then butt welded, and the weld ground flush.

A typical application places the metallic O-ring in axial compression between parallel faces which are square to the fluid passage or vessel axis. The seal is usually located in a conterbore or closed groove on one face, or in a retainer, which eliminates the need for machining a groove.

Operation

When compressed, the seal tubing buckles slightly, resulting in two contact areas on each seal face, and maximum contact stress between the seal and the mating faces (Fig. 56). The seal allows for some deviation in the sealed faces.

ENGINE CYLINDER HEAD GASKETS

Because engine cylinder heads are such an important use of gaskets, we will cover them separately in this section.

Types of Head Gaskets

The engine cylinder head is retained on the block by bolts or studs and a gasket is placed between the mating surfaces to keep a water, oil, and gas tight seal (see Fig. 53).

In early engines, the cylinder head gaskets were made of solid metal, usually soft copper or lead. These metals were ductile enough to conform to the head and block irregularities and served very well at the low combustion pressures of that time.

In later engines, the horsepower and consequently the combustion pressures and temperatures increased. The solid gasket could no longer give an adequate seal.

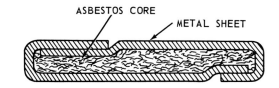

ASBESTOS CORE

METAL SHEET

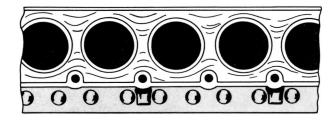

Fig. 57 — Typical Cylinder Head Gasket (Early Engines)

The fabricated gasket was then developed. It was constructed of thin, soft copper sheets with an asbestos core as shown in Fig. 57; however, asbestos is being replaced with other filler material.

This gasket had the resiliency for a tight seal and at the same time absorbed the engine head vibrations.

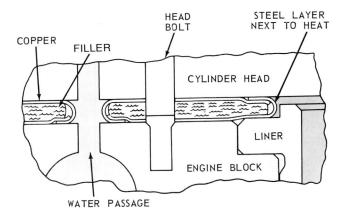

Fig. 58 — Duofold Gasket

However, copper had disadvantages in gaskets. The portion of the gasket exposed to the heat of combustion tended to soften and burn out. To avoid this, tin plated steel was used for the *bottom* layer nearest the heat. Copper grommets were used at the coolant openings to withstand the corrosion.

Later development produced the type of head gasket illustrated in Fig. 58. In this type (Duo-fold), the bottom steel layer overlapped the combustion chamber openings, and the copper layer overlapped the coolant openings.

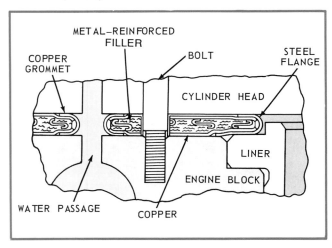

Fig. 59 — Metal-Clad, Metal-Reinforced Gasket

A variation of the layer gasket which has a stronger construction is illustrated in Fig. 59. This became necessary because of the higher pressures in the combustion chambers.

This gasket has a top and bottom layer of copper or steel, with a steel flange at the combustion chamber openings, and copper grommets in the coolant and oil holes. The plain filler (core) is replaced with rein-

forced material. This keeps the filler from squeezing out when the head is tightened to the specified torque. This gasket usually is plastic coated for effective sealing.

All of these cylinder head gaskets are in common use today. However, since each engine demands a cylinder head gasket designed for its particular operation, some engines require more sophisticated gaskets. Some of these are described below.

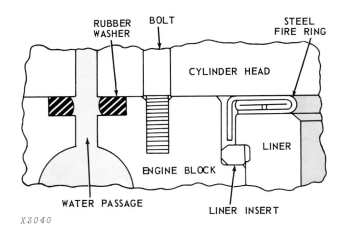

Fig. 60 — Fire Ring And Washer-Type Gasket

The cylinder head gasket design illustrated in Fig. 60 is commonly known as the fire ring and rubber insert type. It is a multi-sealing gasket which has laminated steel fire rings, rubber or neoprene washers, and a rubber or neoprene gasket around the perimeter of the head.

When assembled, the space between the surfaces of the head and block are from 0.002 to 0.010 inch (0.05 to 0.25 mm). The sleeves, water holes, oil holes, and block perimeter are counterbored to accommodate the sealing parts.

The counter bores for the sealing washers are back-tapered from the fluid passages. This keeps the seals from moving into the passages.

The perimeter gasket seals out dirt or other foreign matter and any inner fluid leakage which might develop.

This design is satisfactory. However, one disadvantage is the number of parts. This can be a problem during engine assembly.

This design also requires more accurate machining and stronger structural design of the engine.

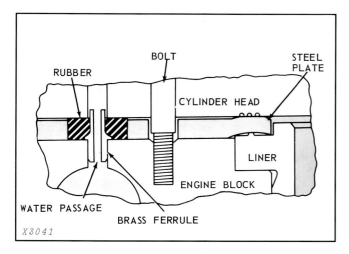

Fig. 61 — Steel Plate Gasket Construction

The steel plate gasket (Fig. 61) is another design of major importance. It consists of a heavy (0.050 to 0.080 inch [1.3 to 2 mm] thick) steel plate assembly. In effect, this is an integral fire ring in a carrier plate with washers and inserts to seal the fluids. The surface of the head and block engage a metal surface. The embossed ring at the cylinder openings keeps a high load on the sleeves to seal the combustion gases. The concentric grooves in the head lock the plate in position and increase the unit loading in this area.

The use of this type sealing requires precise machining of the head and block. Also, an accurate fit of the sleeves in the block is vital to reduce coolant leakage.

Care must also be taken to provide uniform sleeve standup from the block for uniform loading at each of the cylinders with a minimum of head distortion.

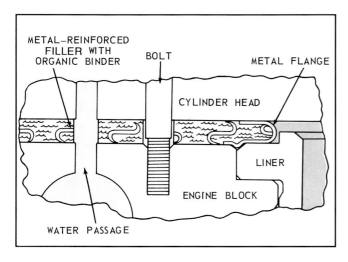

Fig. 62 — Flanged Metal Reinforced Filler Gasket

Fig. 62 illustrates a different concept in head gaskets. This design has metal flanges for the combustion chamber openings. The body of the gasket consists of filler facing material on either side of a perforated steel core. This allows the combustion chamber flange to embed itself into the gasket body, providing contact and sealing of the coolant and oil openings in the rest of the gasket.

The filler surfaces are coated with antistick compound to aid in removal at disassembly.

When this type of gasket is retorqued, its compressibility is reduced. The gasket becomes denser and reaches a more stable condition. Beyond this point, loss of torque is negligible and the tendency to extrude is minimal. In effect, after retorquing, this gasket has the same advantages as the steel plate gasket and face to face design.

The composition and structure of the filler body eliminates the need for washers or inserts in oil or coolant openings; the use of metal grommets is limited to isolated holes.

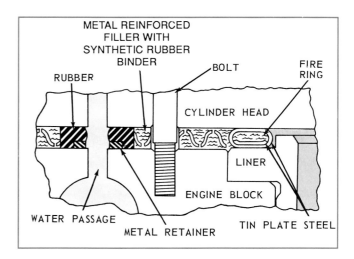

Fig. 63 — Fire Ring And Compressible Plate Gasket

A later gasket concept is illustrated in Fig. 63. This design has all of the advantages and features of the preceding ones.

It has separate fire rings, with filler cores, and a filler "plate" gasket. All of these are compressible.

The individual fire rings are constructed with a fold over flange having a core of multiple layers of steel and filler washers in suitable combination.

The plate gasket itself can seal the coolant and oil, or it can have washers or inserts at the various openings.

The plate section of the gasket consists of synthetic rubber bound facings on either side of a perforated core. The surfaces have an antistick coating.

The individual fire rings with the filler core can compensate for differences in sleeve standup from cylinder to cylinder.

A current head gasket design is shown in Fig. 64. This gasket contains a separate fire ring protected by a stainless steel plate nearest the combustion chamber openings.

The gasket consists of a facing material with a steel core for reinforcement. The facing material surfaces are coated with antistick compound so they are easier to remove.

The composition and structure of the body eliminates the need for washers or inserts in the oil or coolant openings.

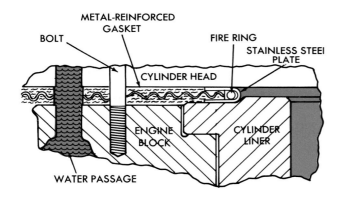

Fig. 64 — Fire Ring Type Gasket

Maintenance Of Head Gaskets

Cylinder head gaskets require care in handling and installation.

When replacing a head gasket that has failed, always try to determine the cause of the failure. This could be be due to not retorquing after installation or improper torque procedure. Proper torquing of head bolts will be discussed later.

Head gasket failures also can be attributed to increased cylinder combustion pressures caused by the installation of superchargers or blowers. Sometimes failure can also be caused by over-fueling in diesel engines to increase horsepower.

Installing Cylinder Head Gaskets

1. Do not use an old head gasket. While it may look good, invisible factors can lead to early failure if it is reused.

2. Be sure that the surface of the engine block and head are scraped and wiped clean to remove all traces of gasket or sealer. Look for scratches and nicks. If necessary, check the surfaces for warp or heat cracks.

3. Be sure decks of head and block are flat. These surfaces may be reclaimed by milling or surface grinding. If this is done, check valves and adjust to correct depth.

4. Install cylinder liner so flange is flush (or a few thousandths of an inch above) the top deck. Make sure counterbores are clean, flat, and free of cracks.

5. To assure proper clamping force when head is connected to block, see that head bolt bosses, hard washers (if used), and the underhead of bolts are flat.

6. Clean head bolt holes in block. Water or oil left in hole could cause the block to crack because of hydraulic pressure created when bolts are tightened.

7. **Be sure that you have the right gasket.** Minor changes are often made in gaskets which, if not compatible with the engine, could cause heating problems. Information on the correct gasket is obtained from the parts catalog. Inspect the gasket for damage. Never use a damaged gasket. As a final check, compare the new gasket with the old one; be sure all holes match.

8. Unless instructed by the engine Technical Manual, **do not** use any sealing medium on the gasket.

9. Position the gasket on the engine block (Fig. 65). Be sure it is fitted over the locating dowels (if present) and that the correct end is to the front of the engine. Some gaskets are symmetrical and will fit either way without damage. But it is possible that some gaskets, even though they fit either way, if installed incorrectly will close off water or oil passages and cause engine damage. Unless otherwise instructed, install the **smooth** side of the gasket to the block of the engine.

NOTE: Always consult the engine Technical Manual for specific instructions.

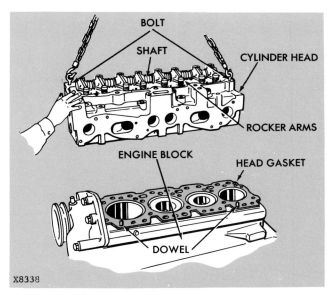

Fig. 65 — Installing Engine Cylinder Head Gasket

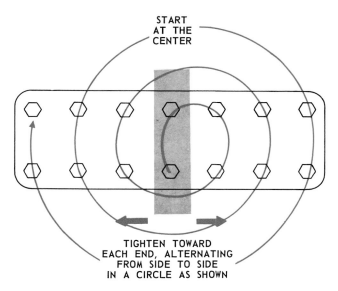

Fig. 66 — Guide To Proper Sequence For Tightening Cylinder Head Cap Screws

10. If the block has no dowels, position gasket by holding fire rings concentric and around outside of the fire dam (not flange) on liners.

11. Carefully install the cylinder head on the block without disturbing the gasket (Fig. 65). Use long bolts with heads cut off to guide head onto block. Use a chain hoist to lower head so it is parallel to block. Be sure the threads of the attaching studs or cap screws are smooth and clean. Coat the threads with oil or white lead for lubrication (if recommended). Install the cap screws or stud nuts and tighten until **just snug.**

12. Tighten each bolt or nut only about one half turn at a time. **Start at the center of the head and work out to both ends of the head until the specified torque is reached.** Fig. 66 shows the general sequence in tightening cylinder heads.

13. After the engine has operated long enough to reach operating temperature, the cylinder head cap screws or nuts should be retorqued. A more effective method is to operate the engine for a specified period. For example, one manufacturer recommends a break in as follows: two hours of break in with the first half hour under no load, the second under half load, and the last hour under 3/4 to full load (dynamometer or otherwise). Then retorque the head bolts.

When retorquing, **always** loosen the bolts before drawing them up to specified torque again. This will insure a more uniform torque on each bolt.

Why Retorque Cylinder Heads?

The real purpose for torquing head bolts is that it provides for a uniform and controlled "squeeze" on the head gasket. When torque is applied to a bolt, some of the turning force is changed to pressure on the gasket, and some of the energy is used up in friction in the threads and under the bolt head. The portion of the turning force which puts a clamping force on the gasket "stretches" the bolt.

Remember, however, that the bolt acts like a spring. The drawing in Fig. 67 illustrates this. Shown are two head bolts of different lengths: 5 inches and 2 1/2 inches (127 and 64 mm). Initially they are torqued *equally,* to say 115 lb-ft (156 N·m). Torqued equally, each bolt will "stretch" equally for *each inch of its length.* But since the 5 inch (127 mm) bolt is twice as long as the 2 1/2-inch (64 mm) bolt, the 5 inch (127 mm) bolt stretches twice as much. For this diameter bolt and for this torque, the stretch is about 0.002 inch (0.05 mm) per inch (25.4 mm) of bolt length. Therefore, the 5 inch (127 mm) bolt stretches 0.010 inch (0.25 mm) (0.002 x 5 = 0.010 (0.002 mm 127 = 0.25 mm) while the 2 1/2-inch (64 mm) bolt stretches 0.005 (0.002 x 2 1/2 = 0.005) (0.002 x 64 = 0.127 mm). The head gasket is now held properly and with uniform forces.

But something happens after the engine is thoroughly warmed up and operated on a dynamometer and under load for about an hour. The head gasket begins to settle in place, "flowing" into the uneven areas on the surface of the cylinder head and block. As it settles in place, the cylinder head moves a little closer to the cylinder block.

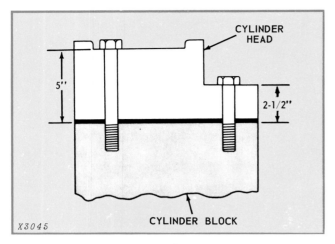

Fig. 67 — Longer Head Bolts "Stretch" More When Torqued

Let's say that the head settles evenly 0.004 of an inch (0.1 mm) (which is why valve clearances must be reset). This settling takes some of the original "stretch" out of the head bolts. To determine the stretch remaining in each bolt, subtract this from the total stretch shown in the chart below.

It should now be clear that the short bolt has lost, relatively, much more of its original "stretch," (and original torque) than the long bolt, so the gasket is now being squeezed considerably less by the short bolts than by the long bolts.

	5-inch Bolt	2½-inch Bolt	(127 mm) Bolt	(64 mm) Bolt
Stretched (torqued)	0.002 inch	0.002 inch	(0.05)	(0.05)
Total Stretch (torqued)	0.10 inch	0.005 inch	(0.25)	(0.13)
Gasket relaxation after warm-up (under load)	0.004 inch	0.004 inch	(0.10)	(0.10)
Stretch remaining	0.006 inch	0.001 inch	(0.15)	(0.25)
Percent of original stretch remaining	60%	20%	60%	20%
Retorque	100%	100%	100%	100%

This problem of bolt "stretch" can never be corrected by **over torquing** head bolts. This only stretches bolts beyond their point of resilience.

Retorquing after a warm up is one practical solution. *(And don't forget to break the bolts loose before retorquing.)* But there is another method of torquing cylinder head bolts that does not require retorquing.

The Torque-Turn Method

Some engine manufacturers have designed the cylinder head, the cylinder head gasket, the cylinder head bolts and the torquing method so that retorquing is not required. There are two methods of Torque-Turn.

• *Line scribe method*

• *Torque angle gauge method*

Line scribe method:

Tighten all cylinder head cap screws in proper sequence (Fig. 66) to specified torque.

Then:

Step A — Make a mark on socket and a second mark 90° (1/4 turn) counterclockwise from first mark on socket. (Fig. 68-A)

Step B — Make a reference mark next to cap screw on cylinder head. (Fig. 68-B)

Step C — Place socket on cap screw so that the first mark on socket aligns with mark on cylinder head. (Fig. 68-C)

Step D — Tighten cap screw 1/4 turn (90°) until second mark on socket aligns with mark on cylinder head. (Fig. 68-D)

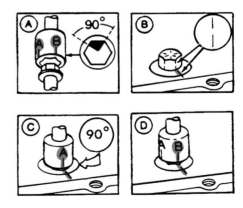

Fig. 68 — Line Scribe Method

Torque angle gauge method:

Tighten all cylinder head cap screws in proper sequence (Fig. 66) to specified torque.

Then follow directions that are provided with the gauge (Fig. 69) and Torque-Turn each cap screw 90°–100° in proper sequence.

NOTE: Retorquing of cylinder head cap screws is not required when this procedure is followed.

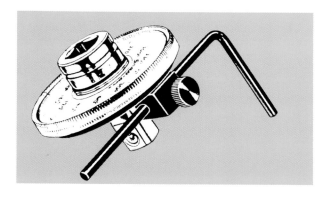

Fig. 69 — Torque Angle Gauge

SEALANTS

Our third major section of "Seals" is **sealants,** which we will cover now.

Sealants are similar to gaskets, except that a sealant is **applied** as a liquid or paste, while a gasket is usually precut to fit and is always placed between mating surfaces.

However, solid *sealing tapes* are also considered "sealants" in this section because they are not cut to size and are also used to cover seams.

Another type of "sealant" is *adhesive,* which is used for its holding power rather than its ability to seal. The functions of sealants are similar to those of gaskets: exclude dust, dirt, and moisture or contain a liquid or gas. However, unlike gaskets, sealants are also used as protective coatings, to reduce noise, to improve appearance, and to perform a joining function.

Sealants are normally used for less severe conditions of temperature and pressure than gaskets (there are exceptions). But sealants are often more resistant to chemicals and are more economical to use than gaskets.

TYPES OF SEALANTS

The major categories of sealants are:

• **Hardening types**
• **Nonhardening types**
• **Tapes**

Whether a sealant is hardening or not depends on its chemical composition and curing characteristics, rather than on its initial form. Sealants generally come in liquid form in a wide range of viscosities. Some epoxy sealers come in powdered form and must be melted when applied. Certain asphalt based sealers and waxes are solid, and are applied by a hot melt technique. Thermosetting film adhesives used for sealing also come in tape form, and they generally require heat and pressure to cure.

Hardening Types

These sealants can be divided into two classes:

• **Rigid—which cure or set up firm**
• **Flexible—which remain flexible after cure**

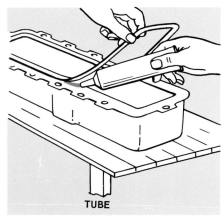
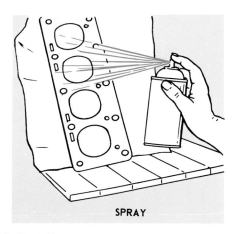

X3046 BRUSH TUBE SPRAY

Fig. 70 — Sealants Are "Gaskets" Which Are Applied In Liquid Or Paste Form

RIGID or hard setting sealants are characterized by their inability to flex. They crack if flexed, and are often difficult to remove. Some, notably the epoxies, can join as well as seal.

Common rigid sealants which are distinctive to this group are those based on components of epoxies, polyesters, acrylics, polyamides, and polyvinyl acetates (PVA).

FLEXIBLE sealants remain flexible after curing, and have elastic bases. Their range of flexibility varies considerably, as does hardness. Some of these sealants are true rubbers; many have good adhesive qualities; and all can be compounded to resist a variety of environmental conditions.

The flexible sealants are either true elastomers that is, they return to their original shape after being deflected or can deform or stretch without tearing, but do not return to their original shape.

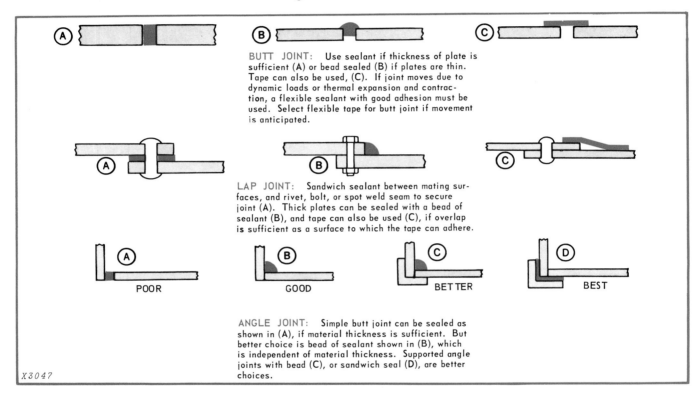

BUTT JOINT: Use sealant if thickness of plate is sufficient (A) or bead sealed (B) if plates are thin. Tape can also be used, (C). If joint moves due to dynamic loads or thermal expansion and contraction, a flexible sealant with good adhesion must be used. Select flexible tape for butt joint if movement is anticipated.

LAP JOINT: Sandwich sealant between mating surfaces, and rivet, bolt, or spot weld seam to secure joint (A). Thick plates can be sealed with a bead of sealant (B), and tape can also be used (C), if overlap is sufficient as a surface to which the tape can adhere.

POOR   GOOD   BETTER   BEST

ANGLE JOINT: Simple butt joint can be sealed as shown in (A), if material thickness is sufficient. But better choice is bead of sealant shown in (B), which is independent of material thickness. Supported angle joints with bead (C), or sandwich seal (D), are better choices.

X3047

Fig. 71 — Common Ways Of Sealing Joints

Nonhardening Types

These soft setting sealants stay "wet" after application and never truly dry. They generally cannot be depended on to perform a joining function, although some formulations are used as adhesives in very low stress joints.

These sealants are characterized by the "mastic" type pastes usually applied to seams with a trowel or brush.

Tapes

Tapes are available in a variety of backings and adhesives usually in pressure sensitive or solvent activated adhesive backs. Self-sticking tapes are also made.

Some nonhardening sealants are formulated so that they can be packaged in "tape" form. These are not adhesive backed and can be easily thumbed into place.

JOINTS FOR SEALANTS

Joints to which a sealant can be applied depend to a great extent on the type of sealant.

For example, a free flowing sealant cannot be applied to a vertical surface; a nonsagging type must be used.

Typical common joints are the simple butt joint, and the lap joint (Fig. 71).

Lap joints can be most easily sealed with tape; fillet sealed, if the thickness of the joined sheets will support a bead of sealant; and sandwich sealed. Sandwich sealing is a common method with structural adhesives. Where ordinary sealants are used, the sealing material is brushed, extruded, troweled, or thumbed on, and the lap joint made and mechanically secured with rivets, or bolts and nuts.

Sealant thickness is more critical when material is applied between surfaces. However, where sealant thickness is not critical, curing time can be delayed considerably by overly thick applications of the material.

If lap joints cannot be sandwich sealed, the joint should have a built in receptacle to receive the sealant.

Voids can result in a joint unless care is taken during application. When beads of sealant are laid in a joint, and air spaces are unavoidable, tooling the sealant before it cures helps to fill the cavities.

When applying sealant, always follow the manufacturer's instructions for proper steps in placing and curing.

USES OF SEALANTS

An inside sealer keeps in a fluid or other medium; an outside sealer keeps out contaminants; some sealers perform both functions.

However, sealants can be used to perform other functions, some of which are almost as important as the primary role of sealing. These are: *(1) Joining; (2) Electrical or thermal insulating; (3) Noise reduction; (4) Vibration dampening; (5) Expansion and contraction control; (6) Smoothing and filleting; (7) Protective coating; (8) Tamper prevention.*

Where conditions are not too severe, sealants can often do the job better and more cheaply than gaskets.

IMPORTANT: Make sure that seals and sealants are compatible with the material they come in contact with. Always follow the manufacturer's recommendations.

TEST YOURSELF

QUESTIONS

1. What are the two **basic** types of seals?

2. Explain the difference between the two basic types of seals.

3. Why should a piston rod seal allow slight leakage of a fluid film?

4. True or false? "As a general rule, replace all seals that are disturbed during repair of a system."

5. How do piston rings seal?
   a. By springing out against the cylinder wall.
   b. By squeezing in against the piston ring groove.
   c. By sealing against the side of the piston ring groove.

6. True or false? "In dynamic uses, the O-ring should be able to roll in its groove."

7. Which are normally used to seal higher pressures?
   a. Metallic gaskets
   b. Nonmetallic gaskets

8. When torquing cylinder head bolts, what is the best sequence?
   a. Start at each end and work to the center.
   b. Start at the center and work toward both ends.
   c. Start at one side and work across.

9. Why must all cylinder head bolts be tightened to the same torque?

10. How do sealants basically differ from gaskets?

11. What is the Torque-Turn method?

(Answers on next page.)

ANSWERS TO PART 1 QUESTIONS

1. (Any three of these.) a. Support the moving part, b. reduce friction, c. reduce wear, d. provide a replaceable wear surface.

2. Axial and radial loads.

3. Full film lubrication which is created by hydrodynamic pressure.

4. Improve heat dissipation, insure that bearing remains round and avoid the possibility of bearing movement.

5. Size of bearing, number of rolling elements and type of races.

6. The full bearing, because of its loading slot, cannot take thrust loads as well as the Conrad bearing.

7. The inner race is going to be the press fit on the shaft because normally the press fit race is **pressed** into the rotating part and the push fit race is **pushed** into place.

8. Force the bearing on the shaft by pressing on the outer races. Only press on the race that is the press fit.

9. Fatigue failure.

10. A distorted housing or shaft.

11. First establish end play, then remove or add shims to get the prescribed preload.

ANSWERS TO PART 2 QUESTIONS

1. Dynamic and static.

2. Dynamic seals are used to seal **moving** parts, while static seals are for sealing **fixed** parts.

3. To lubricate and cool the moving parts and to prevent seal wear.

4. True.

5. a. By springing out against the cylinder wall, and c. By sealing against the side of the piston ring groove.

6. True.

7. a. Metallic gaskets.

8. b. Start at the center and work toward both ends.

9. To put a uniform "squeeze" on the head gasket for a good seal.

10. Sealants are applied as a **liquid** or **paste.**

11. It is the method of torquing cylinder head bolts so that retorquing is not necessary.

RECOGNIZE SAFETY INFORMATION

This is the safety-alert symbol (Fig. 1). When you see this symbol on the machine in this manual, or the machine manual, be alert to the potential for personal injury.

Follow recommended precautions and safe operating practices.

Fig. 1 — Safety Alert Symbol

FOLLOW SAFETY INSTRUCTIONS

Carefully read all safety messages in this manual or the machine manual and on your machine safety signs. Keep safety signs in good condition. Replace missing or damaged safety signs.

Unauthorized modifications to the machine may impair the function and/or safety and affect machine life. Always practice safe maintenance by reading and following safety instructions (Fig. 2).

Fig. 2 — Read Safety Instructions

STAY CLEAR OF MOVING PARTS

Never attempt to service a machine that is running. Always turn the machine off and remove the ignition switch key. Failure to do so may lead to serious injury or death (Fig. 3).

Fig. 3 — Moving Parts Are Hazardous

DON'T USE GASOLINE FOR CLEANING

Gasoline is very flammable. Never use it for cleaning bearings, seals and other machine parts. Gasoline containing tetraethyl lead is also a health hazard.

Fig. 4 — Gasoline Is Very Flammable

ALWAYS USE CLEANING SOLVENTS

Instead of gasoline use a commercial cleaning solvent (Fig. 5). But be sure to read the instructions on the product label and follow them. Provide adequate ventilation when using a solvent.

SMALL PRESSURE SPRAY CAN

BULK CONCENTRATE (DILUTE BEFORE APPLICATION)

Fig. 5 — Use Commercial Cleaning Solvents — Not Gasoline

STORE SOLVENTS SAFELY

Some solvents may be flammable. Keep sparks and open flames away. Other solvents may contain poisonous chemicals. Always keep solvents in closed containers (Fig. 6), never in open pans.

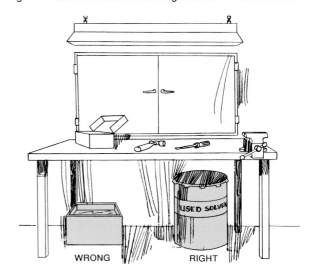

WRONG   RIGHT

Fig. 6 — Store Solvents In Sealed And Labeled Cans, Never In Open Pans

ALWAYS WEAR PROTECTIVE GEAR

Always protect yourself when using solvents (Fig. 7). Wear a face shield, protective gloves and an apron. If you come in direct contact with solvent, rinse it off immediately. Follow directions from the solvent manufacturer.

HARD HAT

FACE SHIELD

RUBBER GLOVES

APRON

CONCENTRATED SOLVENT

Fig. 7 — Wear Face Shield And Rubber Gloves When Cleaning With Cleaning Solutions

PULL BEARINGS CAUTIOUSLY

Removing bearings can be dangerous. Hardened bearing parts can shatter with explosive force. Always wear eye protection (Fig. 8). If possible, cover the bearing area with a shop towel or other protective material to protect yourself and others.

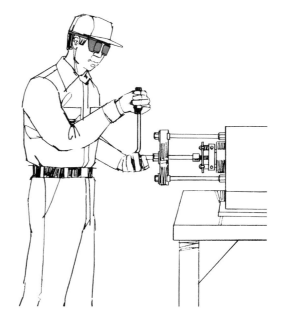

Fig. 8 — Wear Goggles When Pulling Bearings

WEAR EYE PROTECTION

Sometimes bearings and seals have to be driven in or out by hammering. Because bearings are hardened they may shatter and fragments may fly out with significant force. Always wear eye protection (Fig. 9) when driving bearings in or out, and always use the recommended driver for the job.

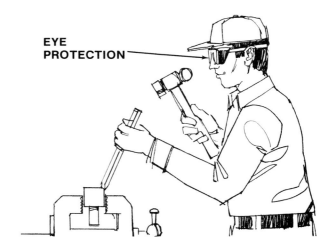

EYE PROTECTION

Fig. 9 — Always Wear Eye Protection When Driving Bearings In Or Out

Litho in U.S.A.

TORQUE VALUES AND METRIC CONVERSIONS / PART 4

UNIFIED INCH BOLT AND CAP SCREW TORQUE VALUES

		1 or 2[b]	5 5.1 5.2	8 8.2
SAE Grade and Head Markings	NO MARK			

		2	5	8
SAE Grade and Nut Markings	NO MARK			

Size	Grade 1				Grade 2[b]				Grade 5, 5.1, or 5.2				Grade 8 or 8.2			
	Lubricated[a]		Dry[a]		Lubricated[a]		Dry[a]		Lubricated[a]		Dry[a]		Lubricated[a]		Dry[a]	
	N·m	lb-ft	N·m	lb-ft	N·m	lb-ft	N·m	lb-ft	N·m	lb-ft	N·m	lb-ft	N·m	lb-ft	N·m	lb-ft
1/4	3.7	2.8	4.7	3.5	6	4.5	7.5	5.5	9.5	7	12	9	13.5	10	17	12.5
5/16	7.7	5.5	10	7	12	9	15	11	20	15	25	18	28	21	35	26
3/8	14	10	17	13	22	16	27	20	35	26	44	33	50	36	63	46
7/16	22	16	28	20	35	26	44	32	55	41	70	52	80	58	100	75
1/2	33	25	42	31	53	39	67	50	85	63	110	80	120	90	150	115
9/16	48	36	60	45	75	56	95	70	125	90	155	115	175	130	225	160
5/8	67	50	85	62	105	78	135	100	170	125	215	160	215	160	300	225
3/4	120	87	150	110	190	140	240	175	300	225	375	280	425	310	550	400
7/8	190	140	240	175	190	140	240	175	490	360	625	450	700	500	875	650
1	290	210	360	270	290	210	360	270	725	540	925	675	1050	750	1300	975
1-1/8	470	300	510	375	470	300	510	375	900	675	1150	850	1450	1075	1850	1350
1-1/4	570	425	725	530	570	425	725	530	1300	950	1650	1200	2050	1500	2600	1950
1-3/8	750	550	950	700	750	550	950	700	1700	1250	2150	1550	2700	2000	3400	2550
1-1/2	1000	725	1250	925	990	725	1250	930	2250	1650	2850	2100	3600	2650	4550	3350

DO NOT use these values if a different torque value or tightening procedure is given for a specific application. Torque values listed are for general use only. Check tightness of fasteners periodically.

Shear bolts are designed to fail under predetermined loads. Always replace shear bolts with identical grade.

Fasteners should be replaced with the same or higher grade. If higher grade fasteners are used, these should only be tightened to the strength of the original.

Make sure fasteners threads are clean and that you properly start thread engagement. This will prevent them from failing when tightening.

Tighten plastic insert or crimped steel-type lock nuts to approximately 50 percent of the dry torque shown in the chart, applied to the nut, not to the bolt head. Tighten toothed or serrated-type lock nuts to the full torque value.

a "Lubricated" means coated with a lubricant such as engine oil, or fasteners with phosphate and oil coatings. "Dry" means plain or zinc plated without any lubrication.

b Grade 2 applies for hex cap screws (not hex bolts) up to 152 mm (6-in.) long. Grade 1 applies for hex cap screws over 152 mm (6-in.) long, and for all other types of bolts and screws of any length.

METRIC BOLT AND CAP SCREW TORQUE VALUES

	Class 4.8				Class 8.8 or 9.8				Class 10.9				Class 12.9			
	Lubricated[a]		Dry[a]		Lubricated[a]		Dry[a]		Lubricated[a]		Dry[a]		Lubricated[a]		Dry[a]	
Size	N·m	lb-ft	N·m	lb-ft	N·m	lb-ft	N·m	lb-ft	N·m	lb-ft	N·m	lb-ft	N·m	lb-ft	N·m	lb-ft
M6	4.8	3.5	6	4.5	9	6.5	11	8.5	13	9.5	17	12	15	11.5	19	14.5
M8	12	8.5	15	11	22	16	28	20	32	24	40	30	37	28	47	35
M10	23	17	29	21	43	32	55	40	63	47	80	60	75	55	95	70
M12	40	29	50	37	75	55	95	70	110	80	140	105	130	95	165	120
M14	63	47	80	60	120	88	150	110	175	130	225	165	205	150	260	190
M16	100	73	125	92	190	140	240	175	275	200	350	225	320	240	400	300
M18	135	100	175	125	260	195	330	250	375	275	475	350	440	325	560	410
M20	190	140	240	180	375	275	475	350	530	400	675	500	625	460	800	580
M22	260	190	330	250	510	375	650	475	725	540	925	675	850	625	1075	800
M24	330	250	425	310	650	475	825	600	925	675	1150	850	1075	800	1350	1000
M27	490	360	625	450	950	700	1200	875	1350	1000	1700	1250	1600	1150	2000	1500
M30	675	490	850	625	1300	950	1650	1200	1850	1350	2300	1700	2150	1600	2700	2000
M33	900	675	1150	850	1750	1300	2200	1650	2500	1850	3150	2350	2900	2150	3700	2750
M36	1150	850	1450	1075	2250	1650	2850	2100	3200	2350	4050	3000	3750	2750	4750	3500

DO NOT use these values if a different torque value or tightening procedure is given for a specific application. Torque values listed are for general use only. Check tightness of fasteners periodically.

Shear bolts are designed to fail under predetermined loads. Always replace shear bolts with identical property class.

Fasteners should be replaced with the same or higher property class. If higher property class fasteners are used, these should only be tightened to the strength of the original.

Make sure fasteners threads are clean and that you properly start thread engagement. This will prevent them from failing when tightening.

Tighten plastic insert or crimped steel-type lock nuts to approximately 50 percent of the dry torque shown in the chart, applied to the nut, not to the bolt head. Tighten toothed or serrated-type lock nuts to the full torque value.

[a] "Lubricated means coated with a lubricant such as engine oil, or fasteners with phosphate and oil coatings. "Dry means plain or zinc plated without any lubrication.

MEASUREMENT CONVERSION CHART

Metric to English

LENGTH
1 millimeter = 0.039 37 inches ....in
1 meter = 3.281 feet ....ft
1 kilometer = 0.621 miles ....mi

AREA
1 meter^2 = 10.76 feet^2 ....ft^2
1 hectare = 2.471 acres ....acre
(hectare = 10,000 m^2)

MASS (WEIGHT)
1 kilogram = 2.205 pounds ....lb
1 tonne (1000 kg) = 1.102 short ton .....sh tn

VOLUME
1 meter^3 = 35.31 foot^3 ....ft^3
1 meter^3 = 1.308 yard^3 ....yd^3
1 meter^3 = 28.38 bushel ....bu
1 liter = 0.028 38 bushel ....bu
1 liter = 1.057 quart ....qt

PRESSURE
1 kilopascal = .145/in^2
(1 bar = 101.325 kilopascals)

STRESS
1 megapascal or
1 newton/millimeter^2 = 145 pound/in^2 (psi) ....psi
(1 N/mm^2 = 1 MPa)

POWER
1 kilowatt = 1.341 horsepower (550 ftlb/s) ..hp
1 watt = 1 Nm/s)

ENERGY (WORK)
1 joule = 0.000 947 8 British Thermal Unit.BTU
(1 J = 1 W s)

FORCE
1 newton = 0.2248 pounds force ...... lb force

TORQUE OR BENDING MOMENT
1 newton meter = 0.7376 pound-foot ....(lb-ft)

TEMPERATURE
$t_C = (t_F - 32)/1.8$

English to Metric

LENGTH
1 inch = 25.4 millimeters ....mm
1 foot = 0.3048 meters ....m
1 yard = .9144 meters ....m
1 mile = 1.608 kilometers ....km

AREA
1 foot^2 = 0.0929 meter^2 ....m^2
1 acre = 0.4047 hectare ....ha
(hectare = 10,000 m^2)

MASS WEIGHT
1 pound = 0.4535 kilograms ....kg
1 ton (2000 lb) = 0.9071 tonnes ....t

VOLUME
1 foot^3 = 0.028 32 meter^3 ....m^3
1 yard^3 = 0.7646 meter^3 ....m^3
1 bushel = 0.035 24 meter^3 ....m^3
1 bushel = 35.24 liter L
1 quart = 0.9464 liter L
1 gallon = 3.785 liters L

PRESSURE
1 pound/inch^2 = 6.895 kilopascals
= 0.06895 bars

STRESS
1 pound/in^2 (psi) = 0.006 895 megapascal ....MPa
or newton/mm^2 ....N/mm^2
(1 N/mm^2 = 1 MPa)

POWER
1 horsepower (550 ftlb/s) = .7457 kilowatt ..kW
(1 watt = 1 Nm/s)

ENERGY (WORK)
1 British Thermal Unit = 1055 joules ....J
(1 J = 1 W s)

FORCE
1 pound = 4.448 newtons ....N

TORQUE OR BENDING MOMENT
1 pound-foot = 1.356 newton-meters ....Nm

TEMPERATURE
$t_F = 1.8 \times t_C + 32$